Plasma Etching in Semiconductor Fabrication

Plasma Technology

Advisory Editor: L. Holland

Vol. 1. Plasma Etching in Semiconductor Fabrication (R.A. Morgan)

Plasma Technology, 1

Plasma Etching in Semiconductor Fabrication

Russ A. Morgan

University of Sussex, Falmer, Brighton, Sussex
and
British Telecom Research Laboratories, Ipswich, U.K.

ELSEVIER

Amsterdam — Oxford — New York — Tokyo 1985

ELSEVIER SCIENCE PUBLISHERS B.V.
1 Molenwerf
P.O. Box 211, 1000 AE Amsterdam, The Netherlands

Distributors for the United States and Canada:

ELSEVIER SCIENCE PUBLISHING COMPANY INC.
52, Vanderbilt Avenue
New York, NY 10017

ISBN 0-444-42419-9 (Vol. 1)
ISBN 0-444-42469-5 (Series)

Printed in The Netherlands

PREFACE

This book reviews and reports on research studies of plasma etching used in the fabrication of semiconductor devices. The etching characteristics of materials widely used in device fabrication, such as silicon, silicon dioxide, photoresist and aluminium films, have been given most attention. The work is based on the original thesis submitted to the University of Sussex for a post-graduate degree and is presented as a book for more widespread interest. The interest in plasma etching in device fabrication is motivated by the ability to attain anisotropic etching from the ion bombardment component of a plasma. This anisotropic etching capability means that surface material is removed without undercutting of the mask layer, permitting line-width fidelity. This can lead to rapid progress in the miniaturisation of device features.

An extensive review is made of etching mechanisms in halogenated plasmas and of reactor architecture. An investigation was made of the ionization conditions existing in one type of planar reactor employed for anisotropic etching of high density surface arrays. This was carried out using Langmuir probes for the determination of ion density, electron temperature, and plasma and floating potentials. Plasma potentials could not be measured because the large electron current collected melted the probe tip. However, measurements were made to determine the applied r.f. potential/current relationship and the negative bias developed at the r.f. electrode.

In addition to this work, a study was made into methods of reducing photoresist mask erosion in plasma etching processes, a necessity in maintaining dimensional control. Most methods consisted of pre-treatments inducing cross-linking in the printed resist pattern. This book provides process information and an introduction to plasma generation, probe theory and photoresist chemistry. Details are given of plasma etching processes and the effects of changes of parameter such as r.f. applied frequency, gas flow, gas pressure and oxygen/hydrogen additions. The author has included a personal view of the subject which arises from the preparation of this thesis. However, this book should afford a student an introduction to the subject, and process engineers will find value in the review of literature. Both students and engineers will find value in the discussion of the plasma conditions in a typical process plant, the photoresist mask erosion study and in the techniques used for characterisation measurements.

The research work contained in this book was carried out at the University of Sussex and at the British Telecom Research Laboratories (BTRL)*. I would like to thank the Directorate of BTRL for making available some excellent facilities for pursuing this research study and in granting permission to use the information. I am grateful to my colleague, M Gill at BTRL, for his vigorous discussions. Above all I wish to express my sincere appreciation of all the advice, guidance and help given to me by Professor L Holland (my supervisor) throughout the period of studies.

R.A. MORGAN

* Author's present affiliation: MMI, 2175 Mission Boulevard, Santa Clara, CA, U.S.

CONTENTS

CHAPTER ONE

INTRODUCTION

1.1 An Example of Masking and Plasma Etching Procedure in Integrated Circuit Production

In the microelectronics industry there is a need for etching processes in the fabrication of devices with fine line width geometries which maintain the same dimensions as the masking layer. This can be achieved with plasma etching, particularly when assisted with ion bombardment normal to the working surface as this has the capability of removing material perpendicular to the surface with negligible lateral removal of material. This is known as 'Anisotropic Etching' and enables accurate mask pattern transfer to be obtained. Typical silicon gate MOS technologies are based on linewidths of 5 μm (1 μm = Micrometer = 10^{-6} metre). A large MOS RAM circuit of 64 Kbit capacity as shown in Fig. 1.1, requires critical dimensions of 3 μm. The capability of producing fine line geometries allows the circuit layout designer to pack more devices per single chip area. Thus given devices can now be manufactured in smaller sizes or more complex devices produced.

The initial step of mask preparation down to a few μm in dimension has depended on the development of processes for producing fine line patterns and this has led to UV photolithography and electron beam lithographic techniques with which it is possible to generate a line width pattern of less than 1 μm dimension in organic polymer films.

Larger geometries can use conventional photoresist development to provide the masking layer followed by wet etching of the electronic

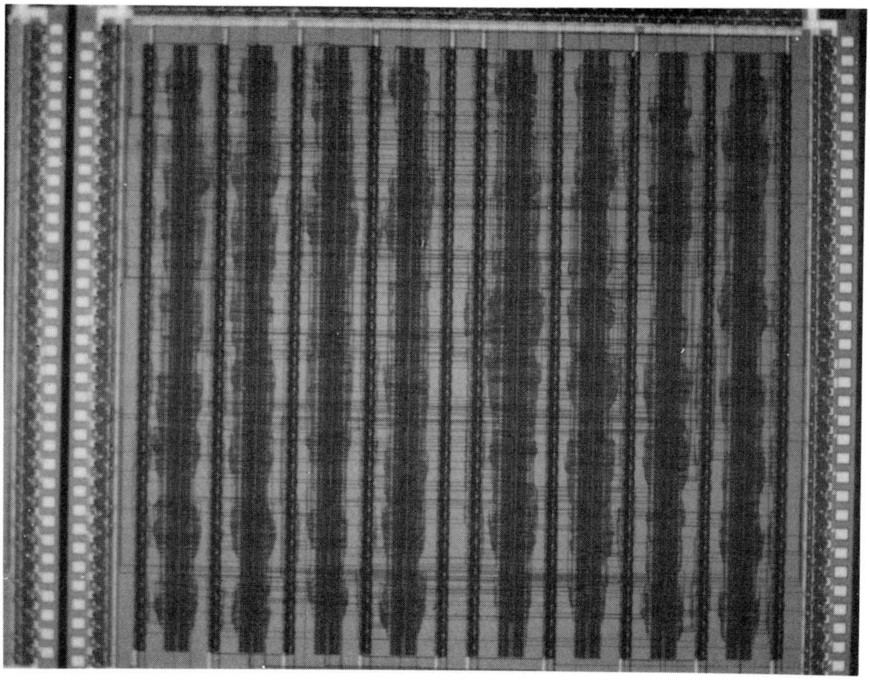

Fig. 1.1. 64 kBit MOS RAM, magnification 50X. The photograph shows a 64 kBit MOS RAM which requires the accurate fabrication on a chip of lines of 3 μm width to achieve high packing density. The magnification is too low for the lines to be seen.

device. However these simple techniques have a number of disadvantages which become more evident as the line width dimensions of the devices are reduced. Conventional resist development can be used from line width down to 2 μm, but generally wet etching cannot be used below

5 μm width for the following reasons:

i) Ragged edges form in the etched material and there is poor dimensional control where the resist is not adherent to the substrate.

ii) Bubbles grow during the etching of the exposed substrate surface thereby acting as localized masks resulting in inadequate substrate removal.

iii) Wet etching removes substrate material both perpendicular to the surface and laterally, commonly known as Isotropic Etching. The lateral etching component results in substrate material being removed from under the resist mask layer so that the line width loss is difficult to control.

Using specially developed plasma techniques it is possible to etch small geometries with minimal undercutting maintaining accurate dimensional control of the etched features. Plasma etching also avoids the previous problems of bubble formation and loss of adhesion of the resist masking layer and in some processes the rinsing and drying operation of the wafers can be eliminated, thus improving throughput.

Silicon nitride (Si_3N_4) films have been used to define the required pattern for oxide growth on a silicon substrate and this is a classic example of the process advantages using plasma etching techniques. Silicon nitride films are deposited by chemical vapour deposition or a plasma technique onto a silicon wafer, which has a thin grown oxide layer on its surface to reduce the stress imposed by the Si_3N_4 on the wafer. The Si_3N_4 is patterned with a photoresist mask and the exposed Si_3N_4 is etched in a CF_4 plasma. Removing the exposed Si_3N_4 with conventional wet etching techniques is difficult because the etchant (phosphoric acid at 180 ^{o}C) degrades the photoresist and destroys the photoresist adhesion to the Si_3N_4 surface.

Thus for the wet etching process it is necessary to deposit silica on the Si_3N_4 film and pattern the silica using a photoresist film, then etch the exposed silica areas. The silica can then be used to define the required pattern in the Si_3N_4 film because the silica mask can withstand the wet etchant.

By using a CF_4 plasma the additional silica mask and the associated process steps are unnecessary as the photoresist mask can be used as a masking material alone during the plasma etching of Si_3N_4. After plasma etching the Si_3N_4 the remaining photoresist is removed and areas of the underlying silicon exposed are thermally oxidized to produce silicon dioxide. The Si_3N_4 is used to mask areas during the thermal oxidation of the exposed silicon and the localized growth of silicon dioxide is used in device fabrication as a mask for diffusion processes or/and as a dielectric layer.

Fig. 1.2 shows the process steps for wet and dry methods of etching during the silicon nitride stage in device fabrication. Each device manufacturermay have his own particular type of semiconductor materials for use on various topographies and thus the requirements and conditions for plasma etching will differ accordingly. However, usually active gas plasmas can be used to remove patterned areas down to the substrate in a short time with minimal line width loss in the etched layer. There must also be a minimum loss in the exposed substrate material so a good <u>etching selectivity</u> is required between the rate the exposed layer is removed and the rate the substrate is removed. A good selectivity is also required between the layer to be etched and the masking layer (usually photoresist). The plasma gas conditions must be arranged to reduce the erosion of the masking layer to a minimum. Plasmas, however, are a hostile environment to organic photoresist films and one can experience difficulty in reducing the effects of ion bombardment and free radial attack. It is possible to improve the plasma resistance of a resist and an account of my work on resist "hardening" to reduce erosion is given in the thesis. Thus the relative etching rates of the mask, the layer to be removed and the substrate are of prime importance in device fabrication and have been extensively studied in recent years.

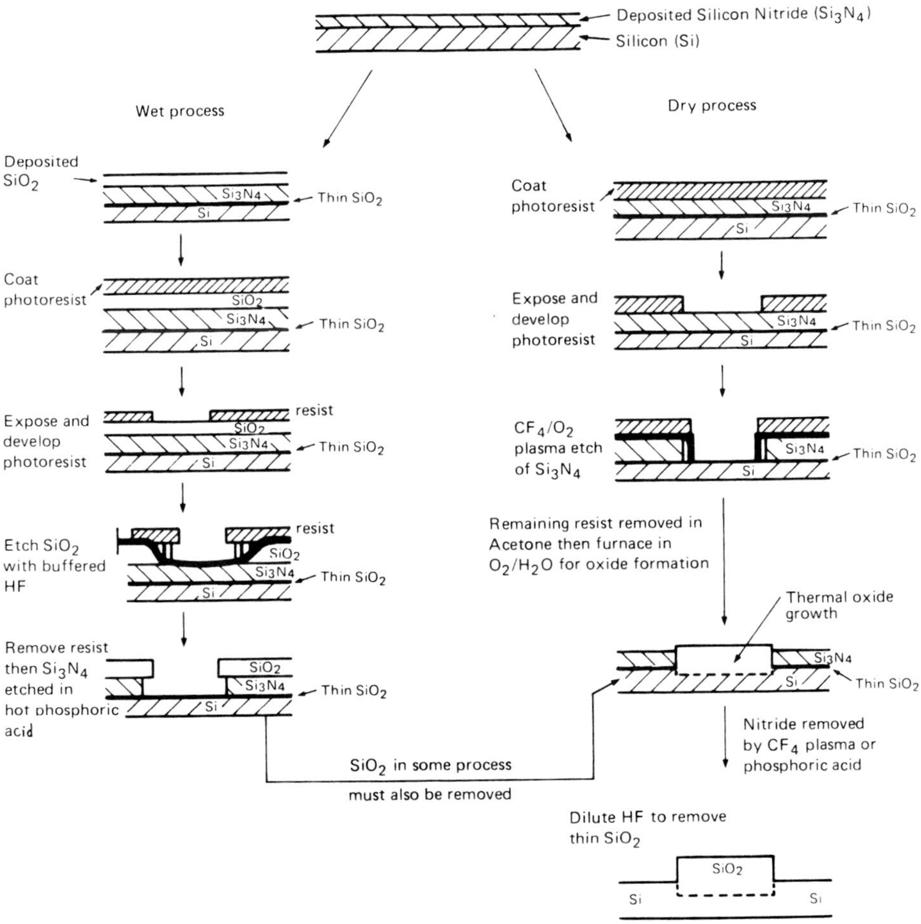

NB. Silicon Nitride is used to mask areas from Silicon oxide growth on Silicon surface

FIG. 1.2 PROCESS STEPS FOR SILICON NITRIDE

1.2 Scope of Study and Experimental Work

Ionized gases are used in several techniques that have been developed for processing semiconductor materials and there have been several reviews on the properties of glow discharges (plasmas) used under various conditions refs. (1-3). These show that a change in the excitation mode and gas pressure can drastically influence the conditions in the ionized gas. For reactive plasma etching gas pressures between 0.01 - 1.00 torr* are used and a particular pressure is usually chosen for optimum performance during the plasma etching process. In this thesis the techniques, processes and possible mechanisms of plasma etching semiconductor materials are initially reviewed and the two principal types of plasma reactor, barrel and planar, are described.

After the review, an account is given of the research undertaken, firstly to determine the plasma conditions in a reactor and secondly to develop a process to reduce resist erosion.

Although plasma reactors are commonly used in semiconductor manufacture, there is little information on such ionization conditions as electron temperature (Te), ion temperature (Ti), plasma potential (Vp) and the nature of the ions interacting with the surface of the semiconductor under treatment. In the study reported here, a flat bed production reactor has been used to investigate the plasma conditions that can typically arise during the etching of aluminium and silicon films.

---

*Although the SI unit for pressure is the Pascal N/m^2 (133.3 Pa = 1 torr) the work in this study is given in torr because so much of the research reviewed uses torr and many vacuum gauges are still calibrated in torr.

This study reports on the diagnostic measurements of electron temperature, ion density, rf voltage and current characteristics. Results are given of the measurement of the dc bias developed by the electrode facing the work surface with respect to ground when inert and active gases are present in the discharge.

Difficulties were experienced when attempting to measure the plasma potential with respect to a coupling electrode with an adjacent ion sheath, because at the high plasma potentials and ion densities existing, the Langmuir probes used were damaged. Thus attention is given to the problems arising when undertaking plasma diagnostic studies in typical process plant.

In addition to the discharge study, experiments are described on methods developed by the writer to cause cross linking in a photoresist film which results in a structure with improved resistance to the erosion effects of the plasma. Photoresists cannot easily be hardened by heating because the materials generally used have a glass transition typically around 130 ^{o}C. Any heating below or at the glass transition temperature produces only a limited degree of cross linking.

The writer has proposed and developed a method of cross linking photoresist without using thermal means. This was achieved by electron beam bombardment inducing bond scission and cross linking with no flow. The resist layer then had an excellent resistance to the erosion effects of the CCl_4 and Cl_2 plasmas. The technique arising from this study has been developed resulting in a patent application*. Several other methods of hardening resist layers for use in plasma etching processes have been investigated and these are also considered here.

---

* R.A. Morgan and C.J. Pollard, Oct. 1982 British and European application. Patent Application No.8213508.

CHAPTER TWO

PLASMA EXCITATION AND REACTOR DESIGN

2.1 Plasma Condition

Low pressure plasmas used in process apparatus are generally characterised by a low degree of gas ionization, (typically 0.1-1%) and an absence of thermal equilibrium between the ionized components, the free electrons and the neutral gas molecules. Thus one can consider three separate temperatures (or separate energy distribution systems). The range of electron energies are between 1 - 20 eV and the ion densities are normally between 10^9 - $10^{12}\,cm^{-3}$.

The mass difference between electrons and ions results in the electrons being accelerated to higher velocities than the ions, in the presence of an electric field. In particular high frequency (hf) and radio frequency (rf) systems the electrons can reach much higher energies than the ions because of their smaller mass and the rapidity of the field change. Ions in the region of the space charge sheath are accelerated towards the electrode surface producing secondary electrons on impact. The secondary electrons are accelerated out into the plasma by the same space charge sheath and may all obtain considerable energy from the sheath.

Ions, because of their larger mass, and smaller mean free path, can more effectively transfer kinetic energy by elastic collision with gas molecules. The average fraction of an electron energy lost in an elastic collision with a gas molecule is $\sim \frac{m}{2M}$ where m is the mass of the electron and M is the mass of the molecule. Thus only 10^{-5} of the kinetic energy is lost by an elastic collision, and the electron energy may continue to rise between collisions.

Typical ratios of ion energy and electron energy in the plasma region of the glow discharge are between 1:10 - 1:1000. This absence of thermal equilibrium makes it possible for the gas temperature (Tg) in the plasma zone to be near ambient, with the ion temperature (Ti) in the plasma region of the glow discharge only slightly higher than Tg, i.e. $Ti \approx Tg$.

However, some of the electrons have enough energy to excite, or ionize molecules or atoms and even rupture molecular bonds during inelastic collisions. Compounds exposed to high temperature (e.g. flames) are composed mainly of radicals and molecules that are incompletely dissociated. However, in glow discharges or plasmas the larger average electron energy produces many more molecular and atomic fragments, which may also be ionized. It is such species produced in an rf discharge with an excitation frequency between 5 kHz and 5.6 GHz which are used in the fabrication of semiconductor devices.

However, as discussed here and in subsequent chapters, an ion sheath may form adjacent to the wafer resting on the electrode surface. Ions from the plasma zone enter the ion sheath and are accelerated towards the electrode, gaining considerable energy. If an inert gas is used the ions are accelerated and bombard the wafer surface removing material by physical sputtering.

Examples of physical sputtering methods to remove material are the rf diode sputter-etcher (4) and the ion beam miller (5).

Physical sputtering methods are generally not applicable to etching silicon and many other semiconductor materials. This is due to the slow sputtering rates and poor selectivity between the layer to be etched and the substrate. However the addition of reactive gases (halogenated hydrocarbon, e.g. CF_4) to a gas discharge can release chemically active gases for surface etching at high rates and increases the selectivity. By careful choice of the plasma chemistry and conditions a volatile compound is formed between the active gas and the material being etched. This volatile compound can be removed from the reaction vessel by pumping.

This is the basis of the plasma etching process, which requires a chamber for the plasma reaction, a flow meter to monitor the rate of gas flow into the chamber and a mechanical vacuum pump to maintain a vacuum in the chamber. The pump should rough out and maintain a pressure in the region 0.01 - 1.0 torr and at the same time remove the reacted compounds. The plasma can be sustained by dc or ac supplies, but it is more usual to use rf excitation because of the charging that occurs on dielectric films when etching in dc plasmas. At low rf frequencies the coupling through insulated layers such as a quartz vessel starts to become difficult and electrodes have to be located within the plasma volume. There may also be other advantages of using rf excitation, such as sustaining plasmas at lower pressures. A typical plasma process system layout is shown in fig. 2.1.

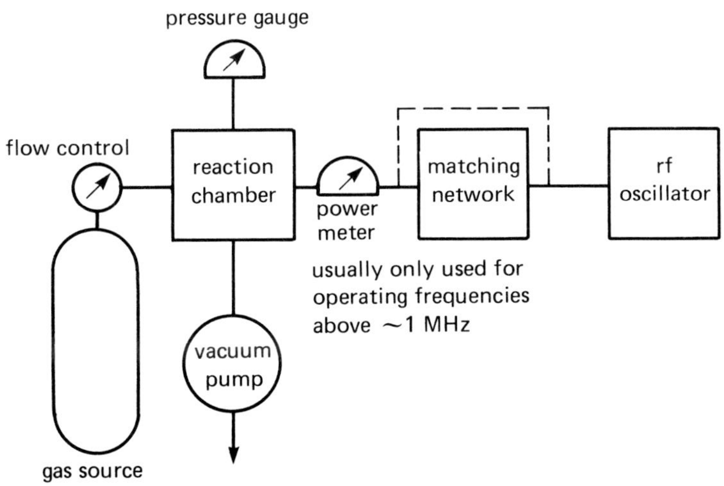

Fig. 2.1.
Typical plasma processing arrangement

2.2 The Barrel Reactor

The barrel reactor was the first type of plasma etcher to be used for semiconductor production purposes and it was derived from an earlier reactor used for plasma oxidation of organic materials, i.e. "Plasma ashing" (6). In this reactor an oxygen plasma at pressures between 2 - 10 torr was sustained by an rf potential applied to two external electrodes. These systems were used to remove photoresist and other carbonaceous materials by oxygen reaction in a plasma. The apparatus was modified for silicon wafer etching by placing an aluminium mesh tunnel around the wafers, which were stacked on long carriers as shown in fig. 2.2. For etching silicon and silicon nitride a typical gas mixture used is carbon tetrafluoride (CF_4) plus 1% oxygen and this mixture is leaked into the chamber to sustain the plasma, which is maintained at between 0.6 - 1 torr pressure.

Barrel reactors typically take between 20 and 50 3 to 4 inch diameter wafers per load thereby giving a high throughput. The mesh of the aluminium tunnel surrounding the wafers is sufficient to allow the passage of activated gas whilst limiting the number of ions passing through. As the rf electrodes are on the outside of the reactor vessel the supply is capacitively coupled to the discharge by the silica wall of the vessel which is a dielectric. Usually the aluminium tunnel is mounted so that it electrically floats. It can be shown that it develops a smaller negative potential with respect to the plasma than the vessel surface. Ojha (169) has studied the potential developed by electrically floating bodies inside metal cages and found that this is lowest when the cages are also floating. Thus in this system the design allows active gas to enter the tunnel from the plasma region with a minimum of energetic ionized components.

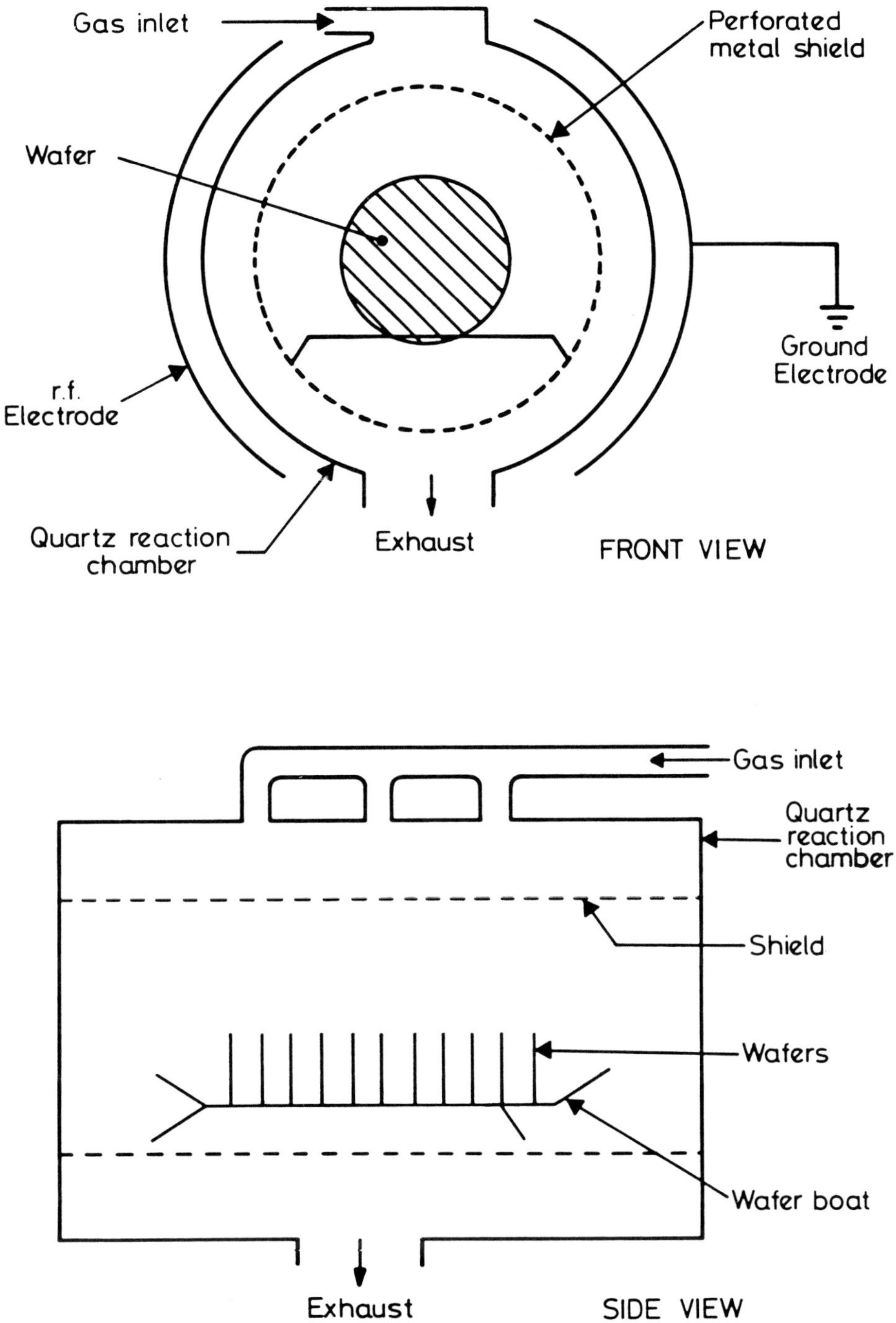

FIG. 2.2
THE BARREL REACTOR.

The wafers are mounted vertically in the reactor chamber and thermal contact is not available for controlling their temperature. Some temperature control is achieved by preheating the wafers in an inert gas plasma before the etching is commenced. Overheating of the wafers gives rise to poor uniformity of etching, poor selectivity in the etching of different layer materials, non-reproducible etching rates and photoresist deterioration. The electron and ion bombardment of a body exposed in the plasma zone can cause undue temperature rise without cooling. The introduction of an aluminium tunnel surrounding the wafers appears to reduce the wafer heating and improves process conditions.

When the aluminium tunnel is used the majority of species arriving at the wafer surface are free radicals, and their concentration is controlled by their recombination rate in the space between the plasma and the wafer surface. The rate of formation of these species is a function of gas pressure and power input, and adjustment of these parameters is necessary to achieve optimum etch rates and maximum selectivity in the etching of different layer materials. The rate of formation of free radical species such as $F^{\bullet}$ and $CF_3^{\bullet}$ will also depend on the reactant concentration CF_4 of their source.

Some typical reactions occurring in CF_4 when excited are as follows:

$$e + CF_4 \rightarrow CF_3^+ + F^{\cdot} + 2e$$
$$e + CF_4 \rightarrow CF_3^{\cdot} + F^{\cdot} + e$$

Formation of active species

$$CF_3^{\cdot} + F^{\cdot} \rightarrow CF_4$$
$$F^{\cdot} + F^{\cdot} \rightarrow F_2$$
$$CF_3^{\cdot} + CF_3^{\cdot} \rightarrow C_2F_6$$

Recombination of active species

Small additions of oxygen can be made to CF_4 plasmas to increase the $F^{\bullet}$ concentration by reducing the $F^{\bullet}$ recombination events. The higher $F^{\bullet}$ concentration will increase the silicon etch rate whilst having very little effect on the silicon dioxide etch rate, this will be discussed in more detail in the next chapter. As etching is mainly by free radicals, which react with the wafers, e.g. $4F^{\bullet} + Si \rightarrow SiF_4$, the process etch time depends on the wafer loading on a particular run.

In this type of reactor where ion bombardment has a small role in the etching process, the etching selectivity for $Si{:}SiO_2$, i.e. the relative etch rates, is as high as 100:1. However barrel reactors suffer a serious disadvantage because, as etching mainly depends on free radical reaction, they etch isotropically and hence cannot be used where line width loss is critical.

2.3 The Planar Reactor

The planar plasma reactor, now widely used for etching was developed from the work of Reinberg (7) and Heineke (8). The system using parallel plate electrodes was an important development because it provided the capability under some conditions for anisotropic etching. A typical reactor consists of a top electrode connected via an insulated feed through to an rf supply and a bottom electrode which is normally grounded, (fig. 2.3). The wafers rest on the ground electrode and are immersed in the plasma. The top electrode is normally capacitively coupled via a blocking capacitor to the rf supply. In this system each electrode is negatively biased to the plasma and generally the negative bias is higher on the top electrode, as shown in the experimental studies.

It is noted here that a planar reactor will etch wafers located on either electrode surface and in some cases will etch wafers faster on the top rf electrode due to the larger negative bias developed on the top electrode. In earlier work, Holland et al.(64) observed enhanced etching of a silica vessel in the regions under the coupling electrodes in an excited vapour of perfluoropolyether. The enhanced etching was due to the larger ion bombardment component in the etched regions.

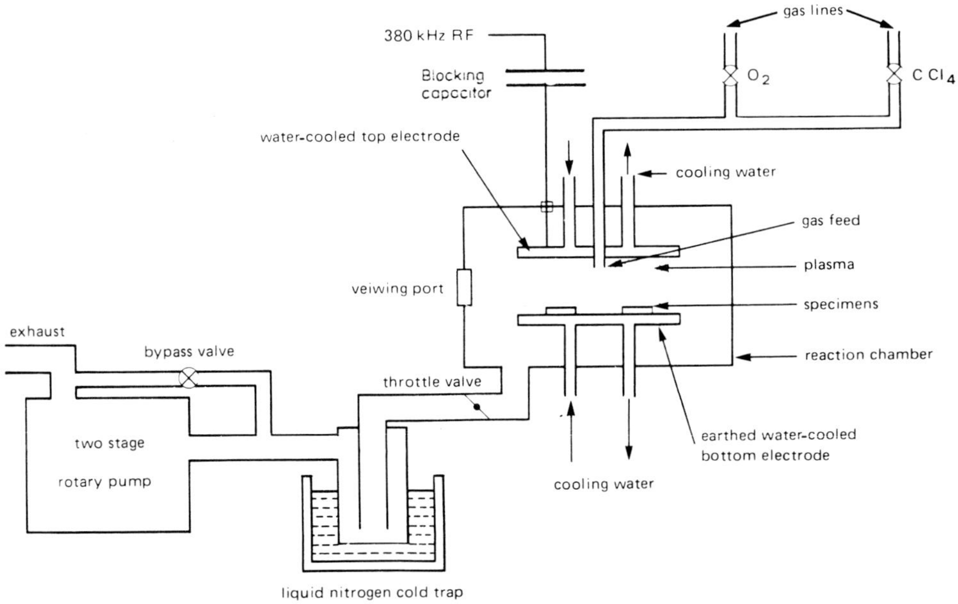

Fig 2.3

A TYPICAL PLANAR REACTION SYSTEM

In the planar system considered here the work surface is the grounded electrode which forms an adjacent ion sheath because of the negative bias developed with respect to the plasma potential. Thus the wafers placed on the ground electrode are bombarded by atomic and molecular ions which have been accelerated in the ion sheath as well as low energy electrons. The wafers are also exposed to free radicals and molecular fragments produced in the plasma volume.

Planar reactors have some disadvantages compared with barrel reactors, e.g. higher cost, poor selectivity, and smaller wafer through-put. The poor selectivity is because of physical sputtering from the ion bombardment contribution which also produces heating effects that can result in poor etching uniformity, rapid removal of photoresist, and resist 'flow' if heating is too excessive.

The first production planar etcher to appear in the semiconductor industry was manufactured in the U.K. by Electrotech and this type of apparatus is the subject of the research study considered in later chapters.

Production planar etchers have water cooled electrodes in order to attempt to improve the uniformity and increase the photoresist life time during etching. A method proposed by the writer to improve resist life which is more successful is discussed in later chapters.

Some machines use chemical resistant pumps which allow the corrosive products to be removed. When wafers coated with aluminium or Al electrodes are etched in a plasma of excited carbon tetrachloride, the volatile product aluminium chloride ($AlCl_3$) is formed and this can rapidly destroy the integrity of pump oils. Some processes use large amounts of oxygen, which can cause explosive problems when pumps with mineral oil lubricants are used. Thus many pumps are designed to use the more inert oils such as those commercially known as 'Fomblin'.

All reactor manufacturers have their own preference for gas mixing in the chamber, gas feed control, electrode material, electrode spacing, chamber size, wafer loading, and rf power supplies and some of these features are shown in figs. 2.4, 2.5 and 2.6. There are now several manufacturers* of planar reactor machines in use, with a wafer

---

*Manufacturers of planar reactors such as Tegal, IPC, MRC, Plasma Therm, Nanotech, LFE and Electrotech.

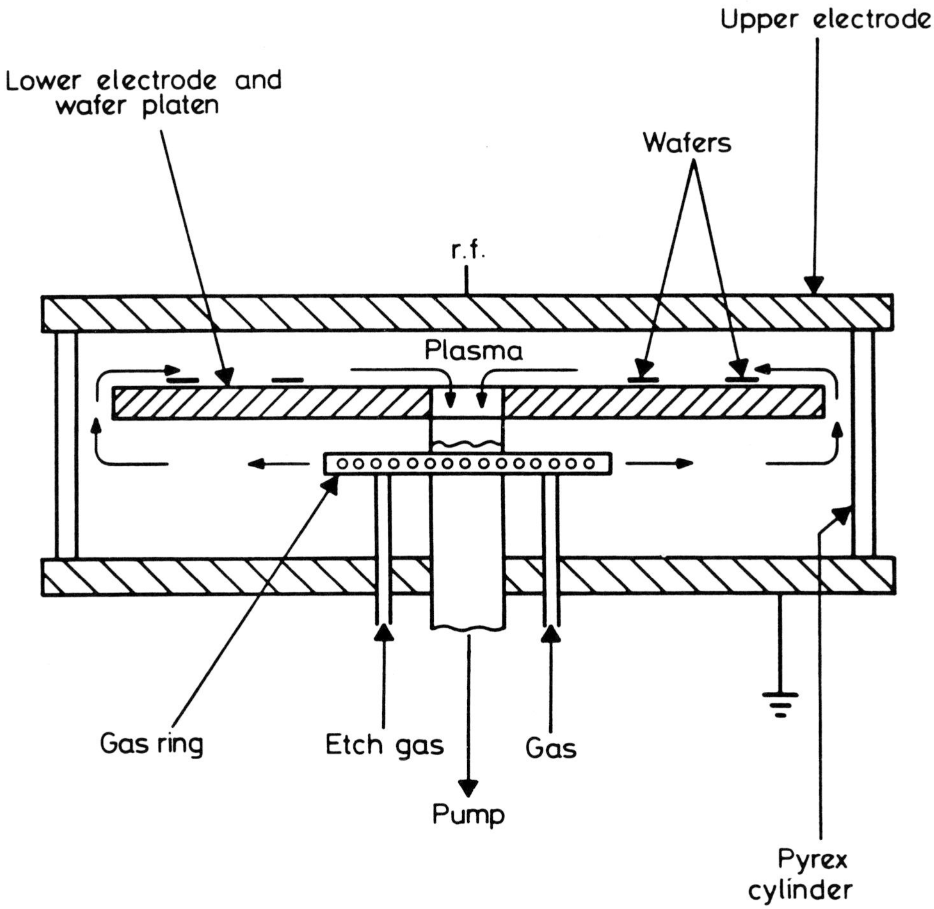

FIG 2.4

REINBERG RADIAL FLOW PLANAR PLASMA REACTOR FROM REF. 17.

throughput varying from that of the research pilot production of 20 wafers per hour to the production system with a throughput of 2000 per hour. The maximum number of wafers being etched per load may also vary from the small research machine of 1 wafer per run holding capacity to the large production machines of 50 wafers per run holding capacity. The machines have facilities available for other etching processes such as removing Al in CCl_4, Si in CF_4 and SiO_2 in C_3F_8. However, in prac-

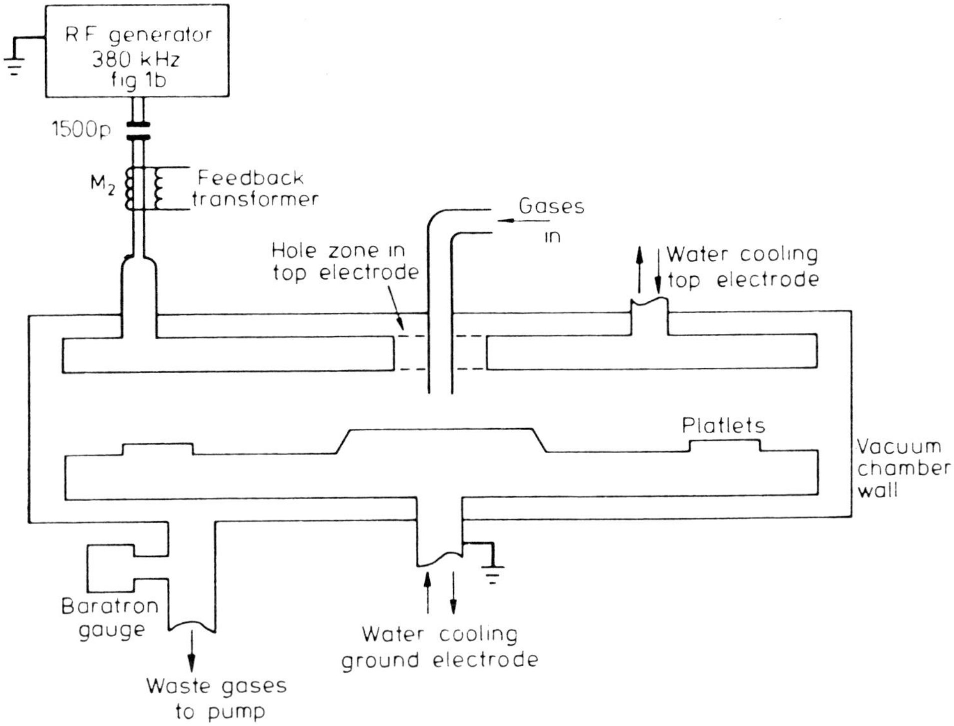

Fig 2.5

Typical Electrotech Planar Reactor

tice, production machines tend to be dedicated to one process because after a specific process the system retains materials which could affect another process.

A problem arising in the large planar reactors is obtaining uniform etching. Thus, if the etching selectivity in a particular process is poor, then serious etching can occur in the substrate of some wafers before others have had the layer to be etched removed. For this reason some manufacturers have made a planar machine for production use, holding and etching a single wafer at a time. These machines have a fast pump down and fast etch rates because the chamber is small and the power density can be high. Although the throughput of these machines

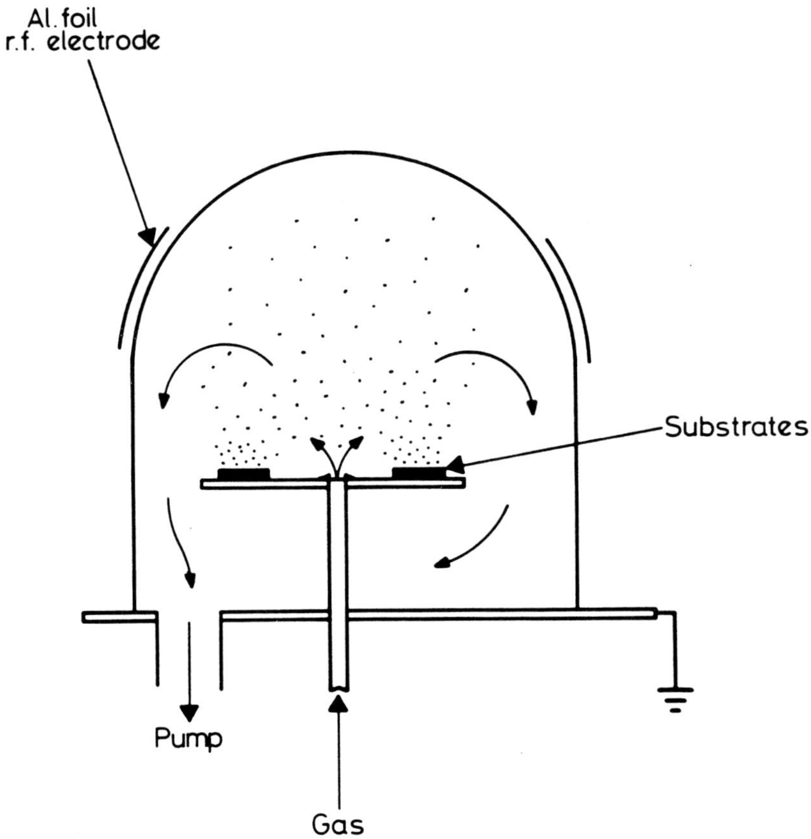

FIG 2.6

HEINEKE'S EARLY EXPERIMENTAL PLANAR PLASMA REACTOR

is obviously smaller than that of large capacity plants they give better uniformity. End point detectors are in use on plasma etching machines to avoid over-etching. End point systems work particularly well on single wafer machines since there is a smaller volume for observation. Also, laser interference techniques can be used to monitor the single wafer in the etcher.

With all plasma process machines it is necessary to precondition the vessel by running the process before the wafers are etched, thus improving etching uniformity and reproducibility. The preconditioning of the chamber (before wafers are placed and etched in the chamber) warms the chamber and reduces the wafer contamination. Winkler (9) has shown by measurement of gas pressure during etching, that the initial pressure rise at the start of the plasma etching process is related to the condition of the chamber and the plasma. The initial pressure rise will be larger for a CCl_4 plasma than an Ar plasma, due to the larger number of fragmentation products produced in the CCl_4 plasma. The final running gas pressures will be slightly different for the same starting pressure for Ar and CCl_4 plasmas when equilibrium is established. This is because the pumping rate will be slightly different when the CCl_4 is partially fragmented in the plasma and the fragmentation products are removed at a different rate from that of the parent molecule.

When etching aluminium in a CCl_4 plasma the reaction product $AlCl_3$ is produced, and some $AlCl_3$ is deposited onto the walls of the chamber. Water vapour entering into the chamber from the atmosphere when loading (or from leaks) reacts with the $AlCl_3$ so that the final wall deposit is the hydroxide. Winkler noted that the initial pressure rise was smaller and reproducible if atmospheric contamination (water vapour) was prevented from entering the plasma vessel when loading the wafers. The lack of water vapour contamination reduced the outgassing in the vessel and thus reduced the initial pressure rise. Prevention of water vapour or atmospheric contamination in the plasma vessel also improved the uniformity and reproducibility of the etching process.

Most plasma reactors used in production are vented to atmospheric pressure for loading using dry nitrogen; and on opening the door to load the wafers into the plasma vessel atmospheric contamination occurs. Winkler prevented atmospheric contamination of the plasma vessel by introducing the wafers through a load lock. The load lock was flushed with dry nitrogen to prevent atmospheric contamination from entering the vessel with the wafer.

The load lock flushed with dry nitrogen allows the plasma vessel door to be opened and so that wafers are introduced from a clean atmosphere. The load lock system with nitrogen flush is a cheaper way than having a load lock which is pre-evacuated before wafer transfer. If an evacuated load lock is used it is possible for the process to have a faster pump down and the plasma can continue to operate whilst loading. Load locks with wafer intake which is automatic are becoming common place in plasma etchers for production processes and this is now beginning to be applied to single wafer plasma etchers.

The gas pressure characteristic not only gives information on the condition of the chamber but also can be used as an end point detector (10). However the most common end point detection systems use optical emission spectra and mass spectroscopy.

The electrode materials used in reactors now tend to be anodised aluminium or titanium. Originally stainless steel and aluminium electrodes were used but aluminium without a thick oxide layer can be reactive and stainless steel can cause stacking faults on the semiconductor wafers, because nickel is sputtered from the electrodes (11) onto the wafer surface. With CCl_4 and CF_4 type plasmas polymer formation has been observed with both stainless steel and aluminium electrodes (12). It should be noted that Fe, Ni and Al are used in the production of stereospecific organic polymers and other derivatives, as Natta, Fischer-Tropsch and Ziegler catalysts. Thus aluminium electrodes are likely to aid the onset of polymerization in the plasma process. However a polymer film can build up over the electrodes surface during processing and reduce the direct contact between the plasma and the electrode material (maintaining an equilibrium for polymer film deposition). Anodized aluminium electrodes present an inert oxide layer to the plasma, but Al_2O_3 can be physically sputtered by positive ions which are accelerated when crossing the space charge sheath adjacent to the electrode. As stated earlier such conditions can exist in planar plasma etchers because they resemble, in principle, rf diode sputtering units.

Vossen (14) has measured a peak to peak applied voltage of 1000 volts for the Electrotech planar etcher and the writer has measured a peak to peak voltage of 950 volts in a similar system. These voltages give rise to high dc bias potential in planar systems which enhance chemical etching of wafer material and also cause physical sputtering of more inert electrode materials. This subject of the plasma etcher characteristic is continued further in later chapters where the potentials developed in a typical system are considered.

Plasma etchers have been designed to operate over a wide rf frequency range, e.g. IPC and Nanotech use a 13.4 MHz supply and Electrotech and Tegal use 380 kHz, 60 kHz and 30 kHz supplies. Thus it is very difficult to compare planar processes, particularly the planar down stream reactor produced by Tokuda Seisakusho which employs microwaves of 2450 MHz frequency to generate the plasma. The plasma is generated in a separate cavity and the reaction products are pumped into a planar type chamber containing the wafers. This is not really considered as a planar reactor since the wafers are not subjected to ion bombardment via a dark space sheath. Comparisons between planar and barrel reactors normally show less line width loss in planar reactors, but they may, as already discussed, give a lower selectivity.

2.4 General Plasma Parameters

Because of the variation between systems, each make of reactor used for a specific etching process will respond slightly differently to a change in plasma gas pressure, power input, gas flow rate, sample loading and chamber temperature.

a. Gas Pressure

Maddox et al. (17) have determined the dependence of the etch rate of SiO_2 in a planar reactor as a function of gas pressure of CF_4 when using different power inputs ranging from 80 watts to 300 watts.

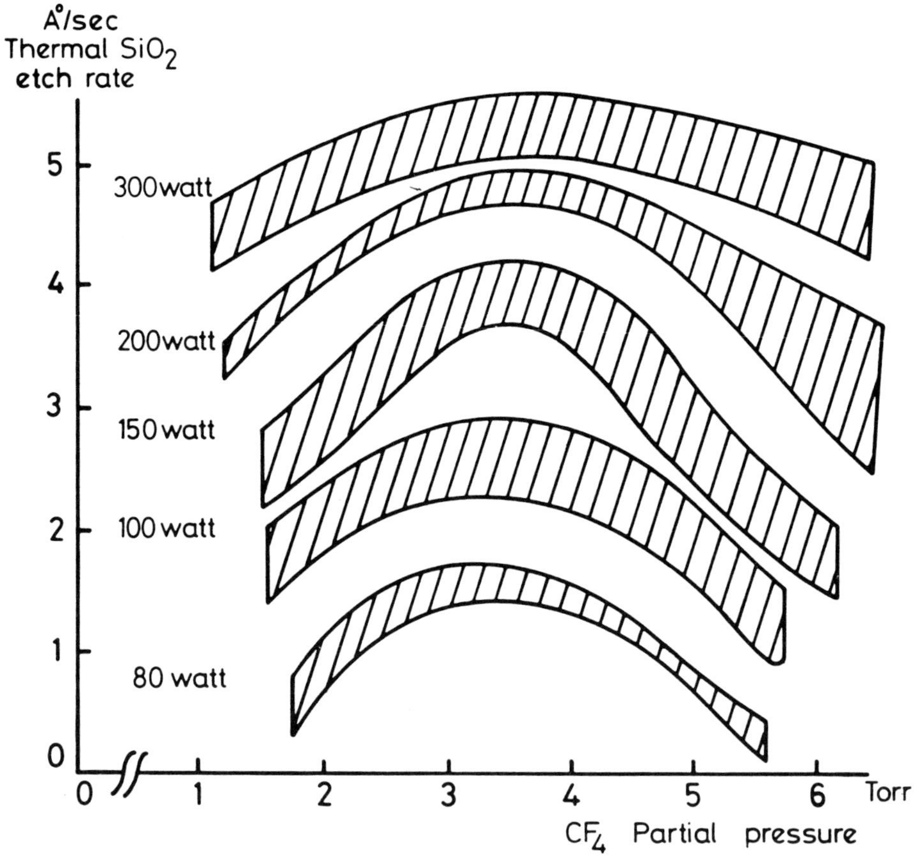

FIG 2.7

SiO_2 PLASMA ETCHING IN CF_4 FROM REF 17.

Their results are displayed in Fig. 2.7. Booth and Heslop (18) have also investigated the etch rate of SiO_2 as a function of gas pressure when using C_3F_8 in an experimental planar reactor and these results are presented in Fig. 2.8.

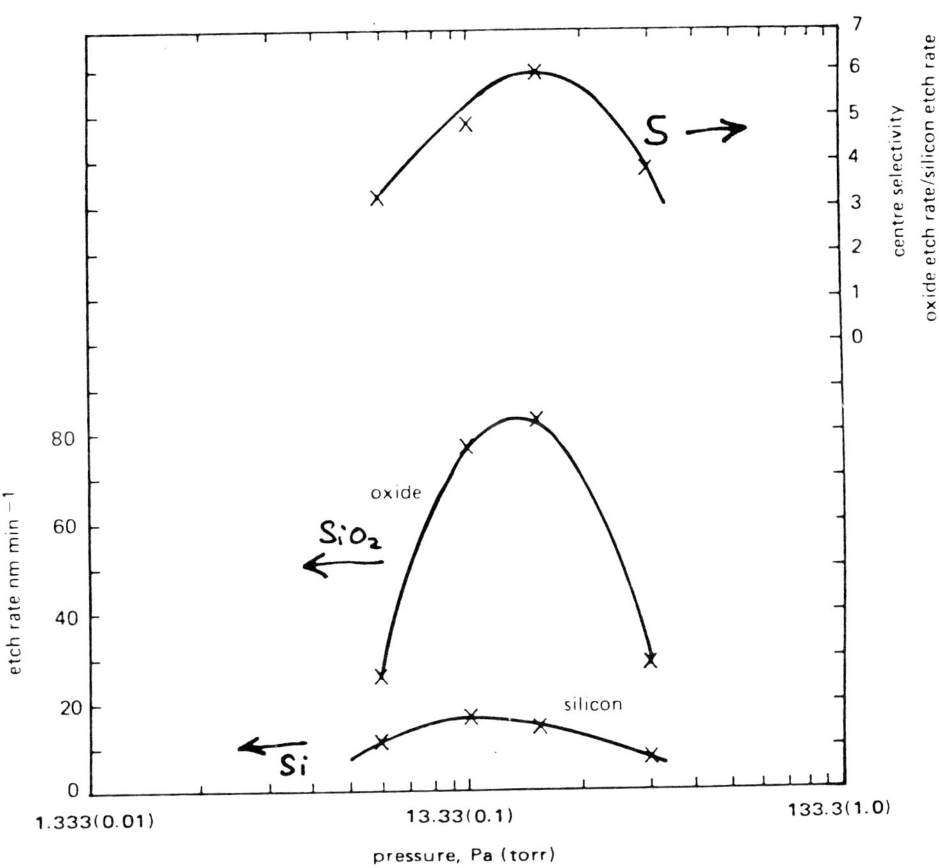

Fig 2.8

Etching characteristics of silicon and thermal oxide in a C_3F_8 plasma (constant power)

The manner in which etching depends on plasma pressure is complex as many effects occur simultaneously. As the gas pressure is raised there is generally a reduction in the mean free path of the plasma components producing more radicals and ions, (depending on the electron temperature and the species in the system). The increase in free radical and ion production can have several effects. For example it can increase the etch rate of the wafer by raising the ion or free radical concentration at the surface of the wafer. On the other hand free radicals can recombine with each other forming higher molecular weight species, e.g. from fragmented CF_4 molecule one can obtain:

$$CF_3^{\cdot} + CF_3^{\cdot} \rightarrow C_2F_6 \text{ (dimer)}.$$

The dimers can also fragment and recombine eventually forming high molecular weight condensed polymers on surfaces. Such an event is more probable at higher pressures, and with long resident times for free radicals of heavy mass.

When increasing the pressure the ion concentration may initially increase and thus the impedance of the plasma may fall thereby resulting in a lower applied potential at a given power input. At the ion sheath the ion bombardment energy will fall with the decrease in sheath potential caused by the reduction of applied potential to the plasma. Thus polymer may grow on an exposed surface if the ions arriving have an energy below the sputtering threshold to remove the polymer from the surface. Also the ion energy may be too low to induce chemical reactions which need ion bombardment.

The increase in ion density at a given power input with a rise in pressure is due to the increase in collision rates between the electrons and gas molecules and between ions and neutrals. The pressure rise will also increase the ionization events and recombination events in the plasma, it may also increase the collisions between ions and neutrals whilst traversing the dark space, thus reducing the ion energy on impact with the electrode surface. Increasing the operating pressures of plasmas may also increase the rate and type of species that could physically or chemically adsorb on the wafer surface. This could then change the rate-controlling step in wafer etching processes.

b. RF Power Input

At constant pressure the plasma density can be raised by increasing the power input into the plasma. This increase in power will cause a rise in the applied potential particularly at lower pressures (15). It will also increase the potential across the ion sheath, thus raising the ion bombardment energy. Any increase in power will also increase the rate of formation of other plasma products such as free radicals. However, any polymer condensation from heavy free radicals will be offset by the higher ion bombardment energy.

In a plasma excited by a low frequency supply the discharge is sustained mainly by ion induced secondary electron emission at the electrodes. An increase in ion bombardment energy will not significantly change the secondary emission coefficient γ_i. Fig. 2.9 shows γ_i for ion impact for an inert gas ion incident on a clean tungsten surface (19). (γ_i is defined as the number of electrons ejected per incident ion.) The curves in Fig. 2.9 indicate that for a particular ion the change in γ_i is small over a large range of energies. The γ_i is more sensitive to changes in ion species and electrode surfaces than the ion energy (19). Thus an increase in ion bombardment energy will not significantly increase the secondary electron emission. However, the larger potential developed across the ion sheath with increasing applied potential will accelerate secondary electrons from the electrode surface out into the plasma with larger energies. The increase in electron energy may raise the ion density and may change the ion: free radical ratios, but will always increase both the ion and free radical

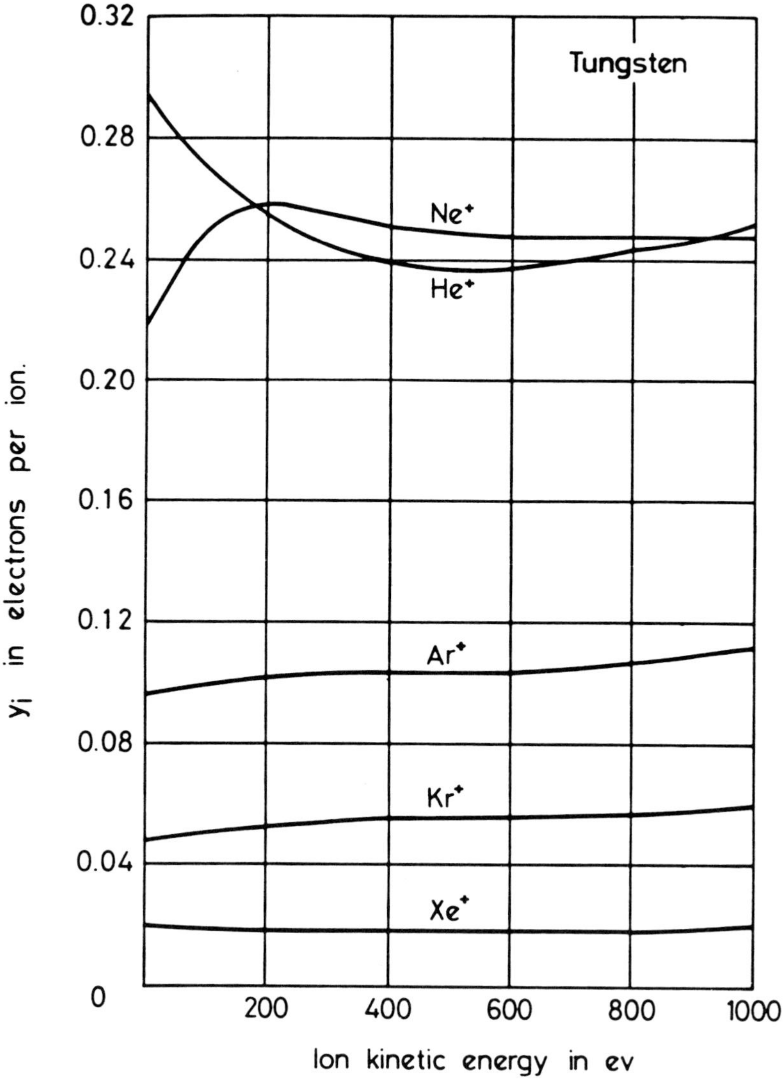

FIG 2.9

y_i FOR SINGLY CHARGED IONS AT DIFFERENT ENERGY FROM REF 19.

concentration. The degree of etch rate rise with increasing rf power input depends on whether the interaction is limited by ion bombardment or free radical species.

The increase in ion bombardment energy by increasing rf power input could make the ion bombardment, at the surface, a more probable reaction for inducing chemical change by bond rupture or damage. There will also be an increase in ion density, which will increase the rate of material being removed from the surface. Alternatively, if the reaction is rate limited by a free radical process, an increase in ion energy or density alone may not be significant. Thus etching will be enhanced by a rise in free radical concentration in the plasma. An increase in ion bombardment energy may also have an effect on surrounding atoms in the area of bombardment which could lower the activation energy and increase the rate of reaction thus increasing the etch rate.

System design may also have an effect on the relation of etch rate with rf power input since the applied rf frequency and type of matching network can vary from one system to another.

The rf power input to the plasma must be correctly controlled because it governs the ion current and is also the major source of thermal energy. In rf plasmas the current flowing in the dark space sheath is by two components conduction, i.e. charge flow and reactive (capacitive) i.e. from cycle changes of the discharges across the sheath.

At 380 kHz the reactive component is less than a quarter of the total current (15), so direct meter measurements can give a fair approximation to the conduction current. RF current measurements to a live electrode can be made using a calibrated ferrite transformer around the power input lead. RF voltages can be measured using a high impedance, low capacitance voltmeter with an rms converter between the rf electrode and ground.

At high frequency $\sim$ 13.5 MHz the reactive components dominate and there are rf radiation losses, thus measurements are usually of power only, by subtraction of forward and reflected power. The power and current controls for a typical planar system are dealt with in more detail by reference 3 and in Chapter 8.

c. Gas Flow

The dependence of gas flow rate on the etching rate of specific wafer surface material varies from plasma reactor to plasma reactor. However, below a critical flow rate, there is an inadequate supply of reactant to etch the wafer material and thus a small increase in flow rate can rapidly increase the wafer etch rate as shown in Fig. 2.10.

As the flow rate is raised the product species are flushed away and thus subjected to less dissociation and recombination reactions in the plasma but at the same time the initial reactant gas is less utilized (27). Use of the etchant gas will increase as the rf power input is increased (27). If the flow rate is further increased with an accompanying increase in pumping speed to give a constant pressure of operation then it is possible that the active species produced in the plasma can be pumped away before they have the opportunity to react. The foregoing effects are illustrated in Fig. 2.11 in relation to the etch rate observed.

The time which a plasma species spends in the system chamber before being pumped is known as the 'residence time' (τ). This is related to the volume of the chamber (V) and the pumping rate S, and is directly related to the flow rate.

$$\tau = \frac{V}{S} \quad \frac{\text{litres}}{\text{litres per sec}}$$

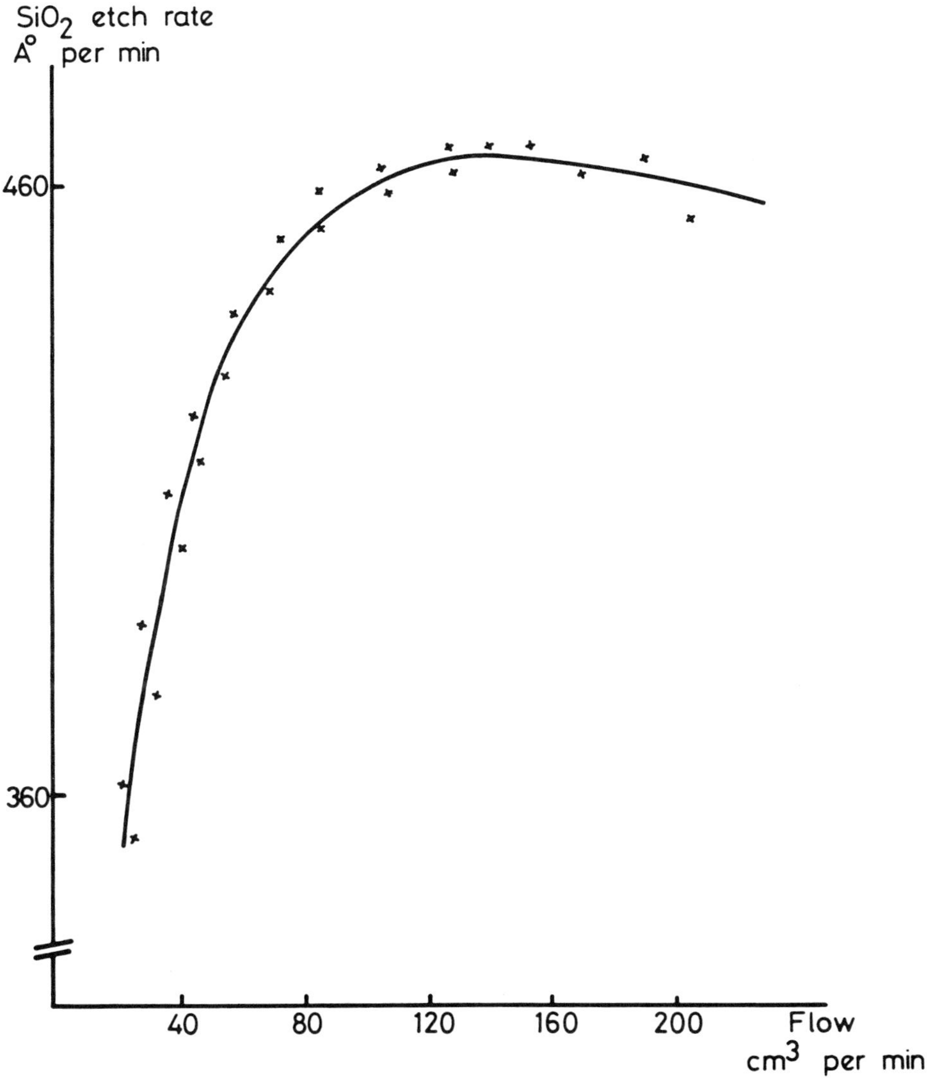

FIG 2.10

SiO_2 ETCH RATE VS CF_4 FLOW RATE AT 100 WATTS POWER 0.5 TORR PRESSURE FROM REF 27.

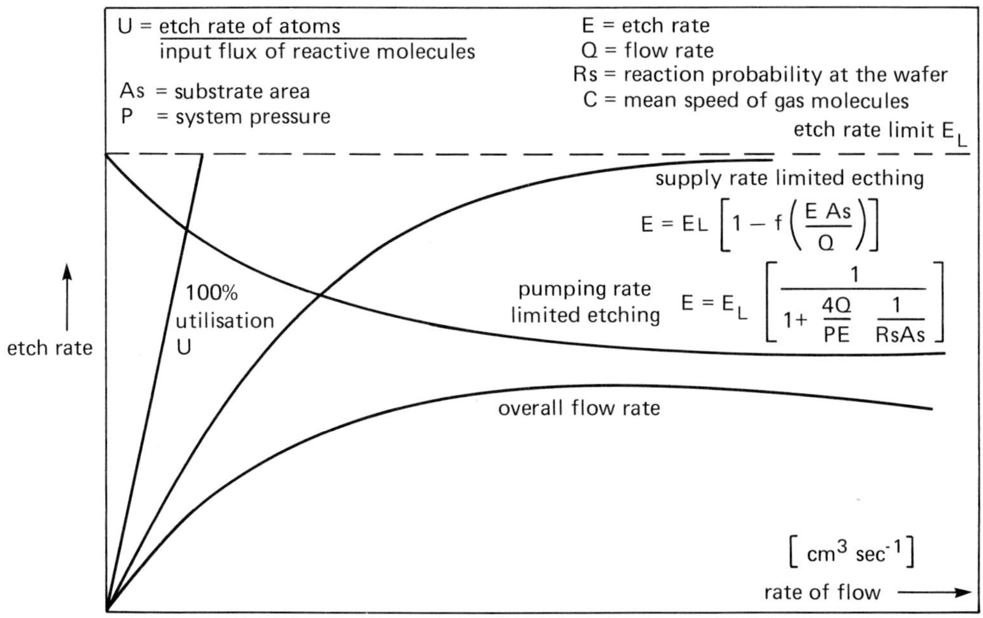

Fig. 2.11 Theoretical model for etch rate vs flow rate from ref 27

The pumping rate is related to the system geometry and pressure, and in many processes the resident time will not be reached by an active species because it is reacting and being consumed thereby producing other products. Residence times for molecules in process systems can vary between 0.5 to 50 seconds providing they are not consumed (28).

d. Loading Effects

Loading effects result from having a limited supply of etching species produced in the plasma. A typical example is the etching of silicon in a barrel reactor in a CF_4/O_2 plasma when the etch rate is reduced by the addition of more silicon wafers. The effect arises from the larger area of silicon present for reaction with the active fluorine species. Maurer et al. (30) have also observed a reduction in the etch rate of silicon in a CF_4 plasma, using a RIE reactor when the silicon wafer load is raised as shown in Fig. 2.12.

Silicon is etched in plasma reactors and reactive ion systems by ion bombardment as well as free radical etching. As the free radical species are more effectively consumed with rise in the wafer load the contribution of the free radical species to the etch rate can fall to a level where ion bombardment is the main component determining the etch rate.

This is an over simplified explanation, as the loading effects can vary depending on the pressure, species present and their resident

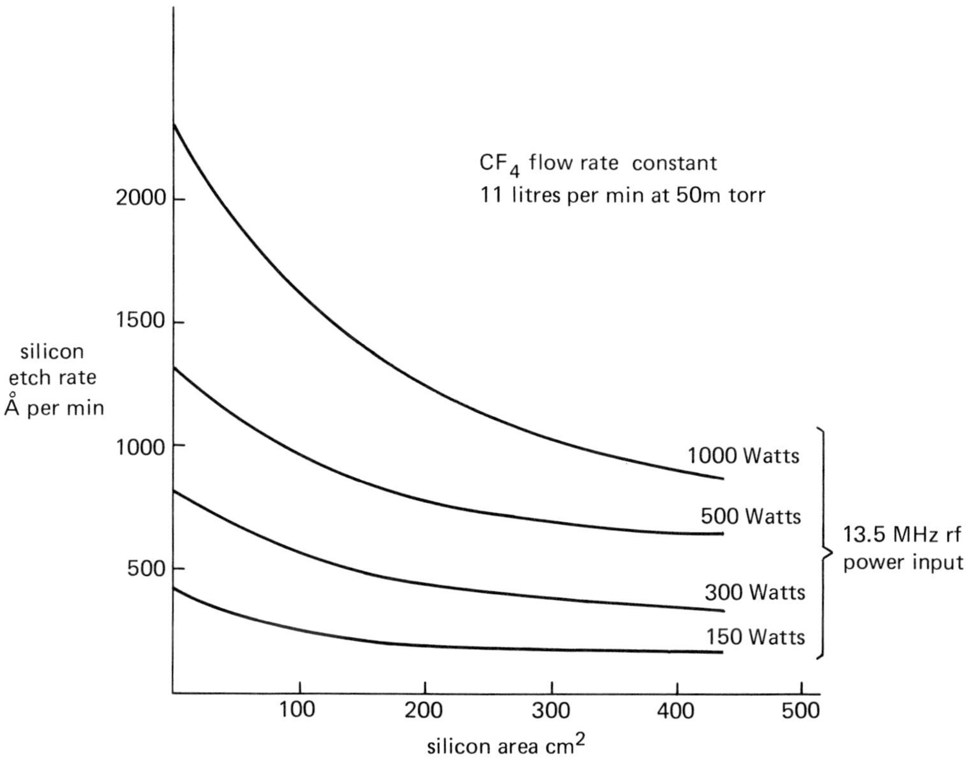

Fig. 2.12 Silicon etch rate in a RIE system vs exposed silicon wafer area to the CF_4 plasma. (30)

time in the plasma. Changes which occur in the gas phase may also cause changes in any surface compound formation which ions may activate or physically sputter. If the active free radicals in the plasma have a high recombination rate and react before they have a chance to etch the wafer then the active free radical concentration will be low. A low concentration of active free radical species in a plasma compared with the ion etching component will result in a smaller loading effect in a system. A low concentration of active free radical species will also reduce the lateral etch component in a plasma etching process resulting in an increase of anisotropy.

e. Temperature

The rate of etching may be dependent on temperature because the chemical reaction involved is subject to an activation energy and also because the surface coverage of chemisorbed species is temperature dependent. Substrate cooling is usually necessary to protect the photoresist mask during etching against thermal deterioration and flow. Substrate cooling can also give consistent etch rates during etching since the wafer temperature can be kept constant.

Poulsen (3) has shown that there is an increase in the etch rate of SiO_2 in CF_4 plasma as the temperature of the SiO_2 sample is increased and his results are given in Fig. 2.13. The shield used in Poulsen's plasma reactor is probably there to remove the ion bombardment component during the etching of the SiO_2 layer thus allowing free radical interaction only; (hence the low etch rates observed in Fig. 2.13).

Most published data on plasma etching involving a large free radical etching component exhibit an Arrhenius type temperature dependence for the etch rate with activation energies lying in the range of

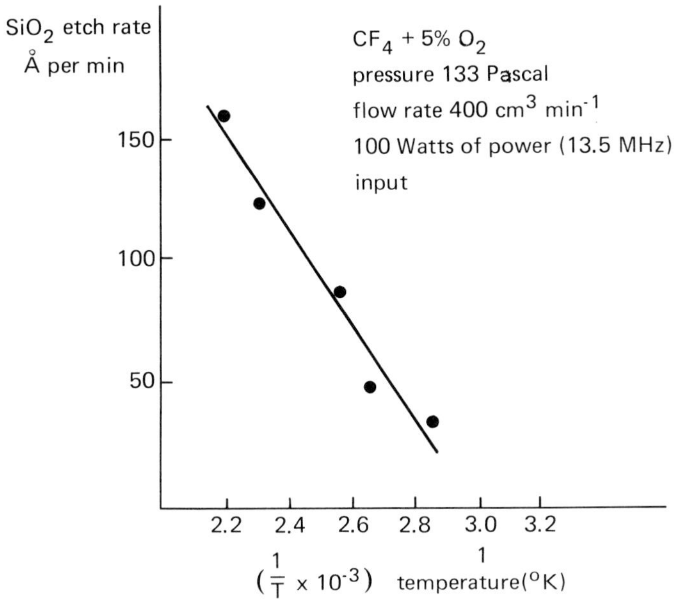

Fig. 2.13.

SiO_2 etch rate vs SiO_2 temperature using a CF_4 plasma with 5% oxygen addition using a barrel reactor with shield. (3)

0.05 - 0.5 eV, examples of which are shown in Table 1. A more convenient form is

$$\log_{10} k = \log_{10} A - \frac{\Delta E}{2.303\ RT}$$

where R. is the universal gas constant $JK^{-1}\ mol^{-1}$ or $eV\ K^{-1}\ mol^{-1}$

T. is the reaction temperature ^{o}K

ΔE is the activation energy $J\ mol^{-1}$ or $eV\ mol^{-1}$

A. is the frequency factor sec^{-1}

K. is the reaction rate sec^{-1}

The activation energy is a measure of the energy barrier which must be crossed in order for the reaction to proceed and the frequency factor is the probability factor for the reaction. Both A and E are independent of temperature to a first approximation.

Table 1

Reactor Type	Material Etched	Etching gas	Activation energy eV/Mole	Refs
Barrel	Photoresist	O_2	0.5	21
Barrel	SiO_2	CF_4	0.19	22
Planar	SiO_2 Si	CF_4+10% O_2 " " "	0.16 0.11	23
Planar	Ti	CF_3Br with additions of He and O_2	0.20	24

The Arrhenius relation is usually written as follows:-

$$K = A\ e^{-\frac{\Delta E}{RT}}$$

The dependence of etch rate on temperature will be less in processes where the etching proceeds mainly by ion bombardment in the presence of active species such as planar reactors, where the ion impact supplies the activation energy.

Holland (1) discusses the etching of carbon in a planar reactor using an oxygen rf plasma. Holland states that ion bombardment removes the carbon surface by chemical sputtering which provides the activation energy to form a volatile carbon monoxide product and that the reaction rate can be increased by oxygen implantation in the carbon surface. Thus in reactive/ion etching processes the effect of temperature will depend on how much wafer etching arises from free radical reactions and how much arises from ion induced reactions.

Some workers (20) have observed a decrease in etch rate with increasing wafer temperature and have suggested that a rise in temperature reduces the number of adsorbed species on the wafer surface which would otherwise be available for etching when activated by ion bombardment.

Flamm (25) using Arrhenius plots of the silicon and silicon dioxide etching rate in atomic fluorine has measured activation energies for Si and SiO_2 which he found to be 0.108 eV mol^{-1} and 0.163 eV mol^{-1} respectively. The atomic fluorine was produced in a molecular fluorine plasma from which Flamm also estimated the ratio of Si to SiO_2 etching by fluorine atoms to be

$$(4.74 \pm 0.49)e^{-\frac{0.055\ eV}{RT}}$$

The values of activation energy may vary from one investigation to another. This is probably due to the difficulty in measuring small activation energies in a narrow temperature range. For silicon this effect may also depend on the orientation (100; 111; 110), and different sample surface areas being used giving rise to loading effects.

Much of the rf power dissipated in the plasma ultimately appears as heat and is the major source of temperature rise (26). In planar reactors, the electrodes should be water cooled and have good thermal contact with wafers thus providing temperature control, reproducibility in etching and an improvement in uniformity, particularly where there is a large free radical etching component.

2.5 Some Basic Rules for the Design of Planar Reactors

Experience by the writer in using a planar reactor in experiments, discussed in later chapters, and the reported work of others, shows that the following practical arrangements are important for reactor design.

(i) A good vacuum pressure should be achieved ($\sim 10^{-4}$ torr) before gas admission and if necessary an oil diffusion pump should be used to remove residual gases. A load lock operated at vacuum pressure or flushed with dry nitrogen is essential for reproducibility and uniformity in the etching process, preventing atmospheric contamination from entering the chamber.

(ii) If aluminium is to be etched in a CCl_4 plasma then a cryogenic trap is essential to remove any water vapour and the corrosive Al_2Cl_6 which is a product formed in the etching process.

(iii) The planar reactor should be dedicated to a single process and not used for general etching. Because of 'memory effects' one process can be contaminated by the previous process due to surface adsorption and pump back-streaming. Back-streaming contamination from rotary pumps may occur particularly at low operating pressures ($\sim 10^{-2}$ - 10^{-3} torr).

(iv) The chamber should be large enough to allow the spacing between the electrodes to be adjusted and to permit the use of different diameter electrodes. This will allow adjustment of rf potentials and ion sheath voltages. The electrodes should be coated with a material which has a low sputtering rate and is inert to free radical and active ion attack. For example, if aluminium electrodes are used, then their surfaces should be heavily anodized using a non solvent electrolyte which gives non-porous coatings.

(v) The wafer handling mechanism should be capable of transporting wafers through the machine smoothly, i.e. without sliding friction or vibration. Friction with a silicon substrate can produce silicon dust which is deposited on the wafer. Also vibration can disturb any dust present with disastrous effects on etched patterns with small geometries.

(vi) The plasma should be in a stable condition without fluctuation of gas pressure, rf power input or electrode work holder temperature and these should be correctly monitored and controlled. The electrodes should be water cooled with no variation in temperature across them and no localized hot spots.

In some plasma reactors chamber temperature or electrode temperature (3) is kept constant by regulating the rf power input. However, this will change the ionization conditions. For this reason it is better to control the temperature of electrodes and chamber with an external water cooling system, and maintain the rf power into the chamber at a constant level, only changing if different etch rates are required.

Temperature measurements can be made ranging from a simple thermocouple (with plasma off), to a more complex infra-red detector monitoring the electrode surface temperature through a port into the plasma volume.

Pressure control can be achieved by using the electrical output of the vacuum gauge via an electrical control stepping motor to open or close a baffle valve thus increasing or reducing the pump speed.

RF current input can be monitored using a small transformer surrounding the rf power cable. When this is used with an rms-DC converter it gives a direct r.f. current read-out. Stabilization can be achieved by comparing this voltage with a reference voltage using a differential amplifier. The amplifier output is then used to control the triode grid and thus the current output of the rf generator.

Monitoring the applied rf voltage can help in improving the diagnostic measurements to a plasma process, such as end point detection (67) where the voltage appearing on the rf electrode changes when the etching is complete. Other end point detection systems, such as optical or mass spectrometers should be available on plasma etching machines to improve the device yield.

2.6 Reactive Ion Etching (RIE)

A reactive ion etcher is a parallel plate or planar plasma etcher where the wafers are etched on the live rf electrode rather than the ground electrode (31,20). When wafers are placed on a live planar rf electrode, which is usually smaller in surface area than the ground surface area, then the electrode to plasma dc potential is usually higher than that between ground and plasma. Thus the live rf electrode develops a dc negative bias to ground and the ions have 'larger' bombardment energies at the live rf electrode. A typical RIE system, based on the use of an r.f. electrode insulated from ground as the work or wafer support which is smaller in surface area than the ground electrode area, is shown in Fig. 2.14.

Reactive ion etchers usually operate with an active gas in the 10 - 100 mtorr pressure range. Operation at this pressure requires high applied rf voltages in the region of 0.5 kV to above 1 kV with the electrode separation distance increased to increase the ionization path between the electrodes.

Operation at low gas pressures means there is an increase in the ion diffusion processes, this allows the plasma to contact a larger ground surface area than it could at higher pressures. Thus a large dc negative potential with respect to the plasma is developed at the wafer/rf electrode. When the plasma potential to ground is small this dc potential measured to ground can approach the peak rf value. The dc negative bias of the target to ground is a nett quantity. The ions that bombard the wafer and rf electrode can reach several hundred electron volts in energy, which on bombardment of the wafer surface, can rupture

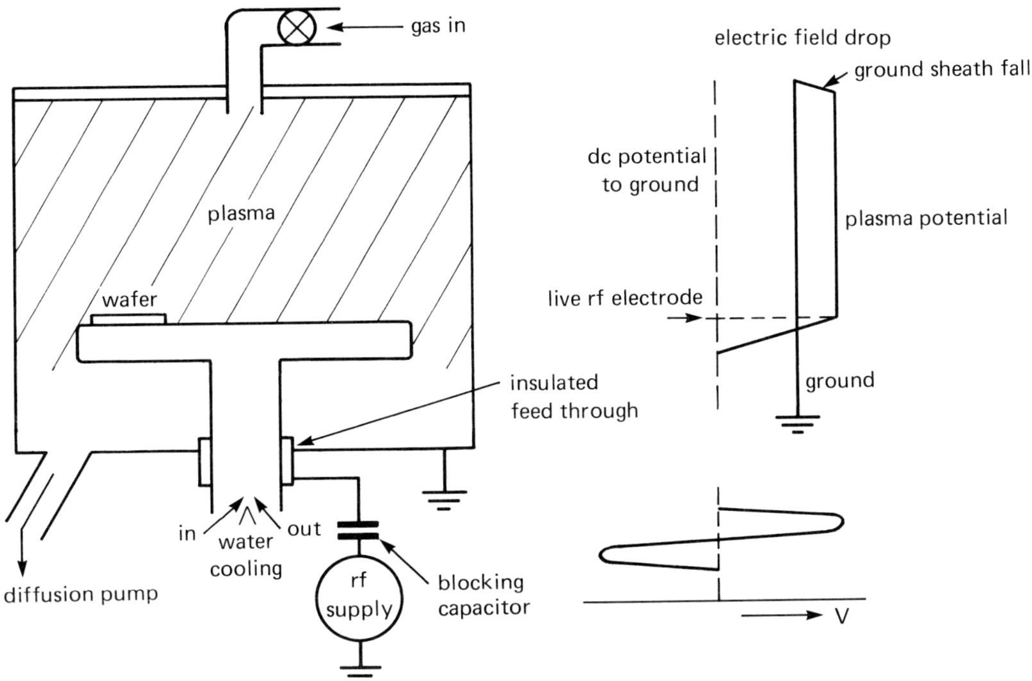

Fig. 2.14 A typical RIE system

bonds and increase the probability of chemical interaction (as well as sputtering).

RIE systems usually have a larger electrode separation than plasma etchers and thus need extra chamber volume. This requires a diffusion pump as well as a rotary pump to achieve the lower gas pressure used. For this reason the wafer throughput of the RIE is usually smaller than that of plasma etchers using the ground electrode as a work surface at higher pressures.

Etching silicon using CF_4 and O_2 mixtures by RIE gives a higher degree of anisotropy than standard plasma etchers using ground work surface. This is because of the lower free radical concentration and the high bombardment energy at the low gas pressures used in RIE systems.

In such low pressure plasmas the ions are not scattered significantly from the normal trajectory in the dark space, consequently a more vertical edge may may be produced. There is, however, a loss in selectivity when using RIE systems (due to the larger ion bombardment energy which is also capable of sputtering) and is therefore not generally used in processes which require good selectivity.

RIE is used for the etching of aluminium silicon copper alloys used as interconnction tracks in integrated circuits. Copper is added to the aluminium silicon alloy to increase the current capability of the interconnection tracks by reducing electron migration. Plasma etching of the Al/Si/Cu alloy using a CCl_4 plasma results in a residue of Cu which is not removed during the etching processes (78). The large ion energy and reduced operating pressure of RIE can be used to sputter away the residual copper as well as chemically etching the Al/Si alloy (86).

As explained earlier RIE can produce ion energies > 500 eV and this is an energy region where ion impact can cause damage in sensitive device areas (33), and may also cause sputtering of the electrodes themselves (14). Bell laboratories,concerned with electrode sputtering particles, have produced the hexode form of RIE shown in Fig. 2.15. This uses vertical work holder rf electrodes facing in different directions to reduce sputtered material landing on the wafer surface.

RIE are made by a number of manufacturers such as Tegal, Plasma Tech, Electrotech, Perkin Elmer, Alvin and IPC. RIE is used where ion bombardment reactions appear to be the rate determining mechanism for etching such as etching contact windows in SiO_2 using, for example, C_3F_8, CHF_3 and CF_4/H_2 mixtures (67). Several reviews have appeared on RIE (34-36) and show concern regarding radiation damage causing neutral trapping centres in devices (which has been investigated using C/V measurements).

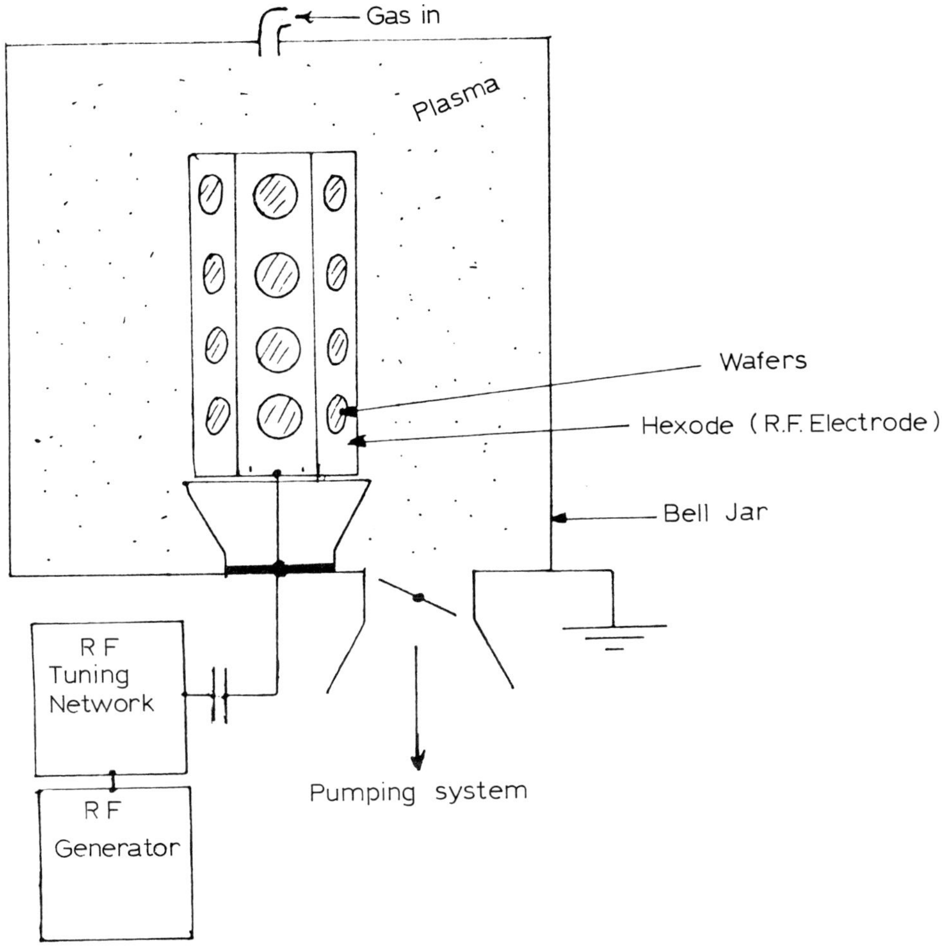

Fig 2.15

Schematic diagram of Bell Laboratories RIE System

2.7 Other Etching Systems

a. The Ion Beam Miller

Ion beam milling is a technique used for surface etching where ions are extracted from an ion source and accelerated in a uniform directional beam onto the substrate surface to be etched (37). Etching in an ion beam miller normally uses an inert energetic beam, such as argon for physical sputter etching (momentum transfer interactions). This is usually done with a substrate or a work holder which can be rotated and off-set with respect to the beam directions. This avoids the growth of surface cones in the etched regions but it imposes limitations on resolution from shadowing which results from the angle of incidence of the beam.

Ion beam milling using a uniform flux of energetic inert ions has many disadvantages compared with planar plasma etchers and RIE systems, with slow etch rates, poor selectivity, trenching, redeposition and very low wafer throughput. If the noble gas is replaced with a reactive gas the energetic beam impinging onto the wafer surface now provides both chemical interaction as well as a physical interaction and is the etching process known as "reactive ion beam etching".

b. Reactive Ion Beam Etching (RIBE)

Reactive ion beam etching systems have several advantages over ion beam milling; they have faster etch rates, better selectivity and no redeposition problems since the etched products are volatile. However its disadvantage compared with plasma and reactive ion etchers is the low wafer throughput. Typical etching pressures for both ion beam milling and RIBE systems are 10^{-5} to 10^{-4} torr which is much lower than that of planar reactors (40).

A type of RIBE that has been used is shown in Fig. 2.16, showing the neutralization filament just below the ion source thus the substrate is bombarded by energetic neutrals and the substrate does not have a charging problem otherwise caused by positive ion bombardment.

In other RIBE systems neutralization occurs at the substrate surface and in some systems neutralized energetic noble atoms such as argon bombard the substrate surface at the same time as a stream of etchant gas such as CF_4 or CCl_4 flows over the substrate surface. The etchant gas is either chemisorbed or physisorbed onto the substrate surface and is dissociated by the arrival of the energetic noble gas atom, producing reactive species and physical disturbance of the substrate in the vicinity of impact thus inducting chemical interaction.

Reactive ion beam etching and ion beam milling is claimed to produce little or no radiation damage (39) and it is suggested that this may be due to the physical isolation of the plasma/ion source thus preventing high energy electrons from bombarding the wafer surface which is thought to be the main cause of radiation damage. Vossen et al. (170)

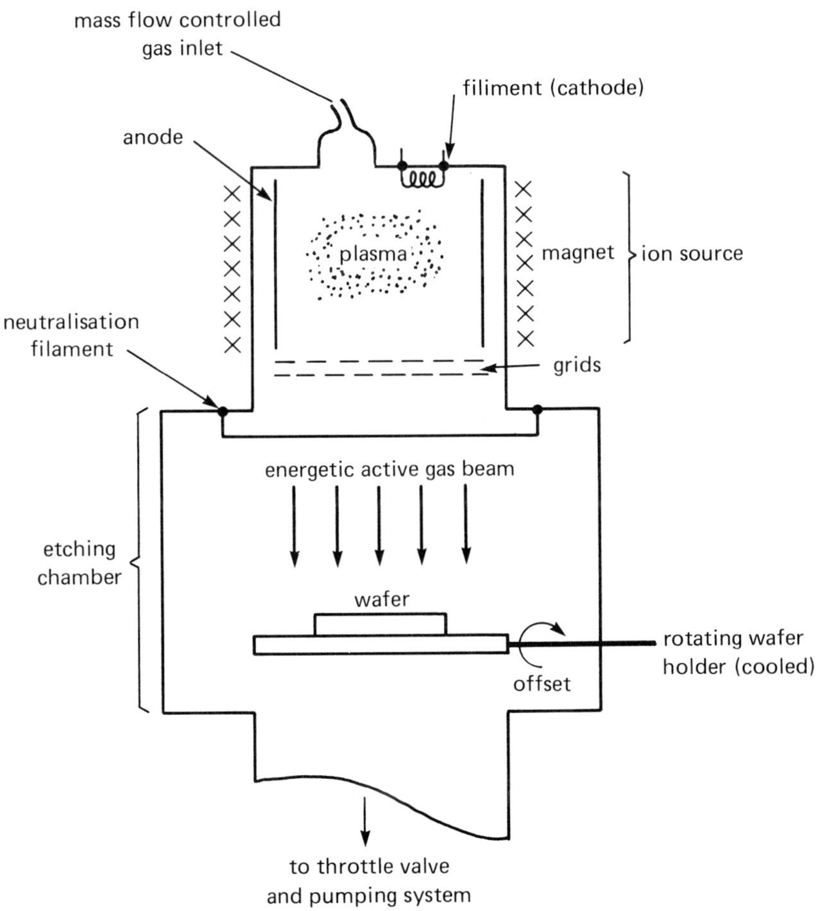

Fig. 2.16 Typical reaction ion beam etcher system using a broad beam ion source,

has suggested that radiation damage may be due to vacuum UV generated by the plasma rather than electron bombardment.

D.F. Downey et al.(39) have obtained good results using RIBE when etching Al/Si/Cu alloy films. They used a CCl_4 plasma and extracted the positive ions (CCl_3^+) obtaining anisotropic etching. By control of the ion beam voltage the copper was also removed by physical sputtering.

It is claimed that etching Al/Si/Cu alloy films with RIBE gave a completely anisotropic etch with a fast etch rate, good selectivity and the remaining alloy does not suffer from post etch corrosion problems. This may be due to free radicals not adsorbing to the side walls of the etched tracks, where they could otherwise absorb water and corrode the alloy tracks.

A commercial machine (Varian RE 580)(39)developed for production RIBE has a 15 cm diameter broad beam ion source giving typical etch rates of aluminium alloy films around 2000Å per min using chlorine. The apparatus has a load lock that allows continuous wafer processing with cassette automatic loading for etching one slice at a time and a pat-erned electrical end point detection system. A summary of typical process conditions, characteristics and particular problems associated with the various types of etchers discussed are shown in Table 2.

Table 2

Summary of typical process conditions, characteristic and particular problems associated with the plasma reactors discussed.

Reactor Type	Process and Conditions Used in Etching	Typical Etch Rates and Selectivities	Problems	Ref
Barrel Reactor IPC type	Etching Silicon Nitride using CF_4+3% O_2 at 100 Watts of 0.7 torr	100 nm per min with 14.1 selectivity over SiO_2 and depends on batch size	Not particularly good selectivity and etches isotropically	18
Planar Reactor Electro Tech 317 Flat bed	Etching Aluminium/Silicon using CCl_4 at 750 Watts 0.3 torr Anodized electrodes	Depends on batch size ~3, 800A per min with 11:1 selectivity over oxide	Will not etch Cu alloys, some lateral etching occurs and resist thickness should be >2μm for 1μm Al etching	18
RIE Experimental Reactor	Etching Si using Cl_2/Ar 2 mtorr and 8 mtorr 0.67 watts cm^2	1000Å per min selectivity 2:1 over oxide but improves with increasing Cl_2 pressure.	Resist erosion Small wafer throughput Poor selectivities	20
Planar Reactor Type 317	Etching Poly Si using Cl_2 0.3 torr 750 watts Anodized electrodes.	1500Å per min selectivity up to 40:1 Si:SiO_2	Although good selectivities Poor uniformity and high resist errorion loss	Work BTRL (67)
	Also using CCl_4	1800Å per min selectivity 5:1 Si: SiO_2 for CCl_4	Poor selectivity Si:SiO_2 and high resist loss	Work BTRL (67)
RIBE Modified Millatron	Etching SiO_2 of 10^{-4} torr ions from C_2F_6 plasma, 1 keV, 0.4 mA cm^{-2}	600Å per min 6:1 SiO_2:Si	Poor etch rate and throughput Poor selectivity	40

ION Miller Results Rock-well	Aluminium etching 0° Argon ion beam 600 eV	400Å per min 1.1:1 for SiO_2:Si	Poor etch rate and throughput Poor selectivity	5

CHAPTER THREE

SILICON AND SILICON DIOXIDE ETCHING IN PLASMA

3.1 General Chemistry

CF_4 gas will not react with silicon or silicon dioxide since the energy requirements are not sufficient for a reaction to proceed. Using thermochemical data, simple calculations can be made to assess if a reaction is energetically favourable to proceed.

For example, the C-F bond strength of dissociation energy in CF_4 is 105 kcal per mole and for Si-Si is 42.2* kcal per mole. If a reaction were to proceed then fluorine would react with the silicon to form the stable volatile product silicon tetrafluoride (SiF_4), which has a high bond dissociation energy of (Si-F) 130 kcal per mole.

The energy balance for the reaction would be:

$$C{-}/{-}F \;+\; Si{-}/{-}Si \;\longrightarrow\; Si{-}F$$

105 kcal 42.2 kcal	130 kcal
bonds broken energy in per mole	bonds formed energy out per mole

Thus the energy requirements to rupture bonds of CF_4 and Si and react to form SiF_4 are too large for a reaction to proceed; hence CF_4 gas will not react with silicon. However, in a CF_4 plasma the high energy electrons collide with molecular bonds and rupture the C-F bond gener-

---

*Si-Si bond formation was obtained from sublimation data.

ating active species such as atomic fluorine and $CF_3\cdot$ radicals as well as ions; e.g.

$$CF_4 + e \rightarrow CF_3^{\cdot} + F^{\cdot} + e$$

$$CF_4 + e \rightarrow CF_3^{+} + F^{\cdot} + 2e$$

CF_4 plasmas are a cheap, simple source for generating active species such as $F^{\cdot}$ and molecular fluorine which will react with silicon to form SiF_4. The energy balance is shown by:

$$2F^{\cdot} + Si\not{-}Si \rightarrow Si{-}F$$

$$42.2 \qquad -2 \times 130 \qquad \Delta H = -217.8 \text{ kcal per mole}$$

(Molecular Fluorine)

$$F\not{-}F + Si\not{-}Si \rightarrow 2Si{-}F$$

$$36.6 \qquad 42.2 \qquad 2 \times 130 \qquad \Delta H = -181.2 \text{ kcal per mole}$$

and for SiO_2, with a bond strength of Si-O, 88.4 kcal per mole

$$F\not{-}F + Si\not{-}O \rightarrow Si{-}F$$

$$36.6 \qquad 88.4 \qquad -130 \qquad \Delta H = -5.0 \text{ kcal per mole}$$

ΔH is the standard enthalpy of formation which is the amount of heat flowing into or out of the reaction system when a compound is formed from the elements under standard conditions (1 mole, latm pressure at 25 ^{o}C). A negative ΔH value indicates that energy is liberated from the system i.e. the reaction is exothermic, with 1 mole of silicon is 181.2 kcal per mole but for 1 mole of F_2 reacting with 1 mole of SiO_2 it is only 5 kcal per mole. Thus it is not surprising that the reaction between F_2 and Si is far more probable than the reaction between F_2 and SiO_2. There is a small increase in $-\Delta H$ when considering SiO_2 reactions with $F^{\cdot}$. Also with the presence of $CF_3^{\cdot}$ radicals the energy favourability may improve still more with the possible formation C = O bonds (173 kcal per mole):

$$\mathrm{F{-}\overset{\displaystyle F}{\underset{\displaystyle F}{C^{\cdot}}} + Si{-}O{-}Si + F^{\cdot} \rightarrow \,>C = O + Si{-}F}$$

$$105 \qquad 2 \times 88.4 \qquad\qquad 173 + 260$$

$$\Delta H = -151.2$$

In thermodynamics ΔH is used to estimate ΔG values, e.g.

$$\Delta G = \Delta H - T\Delta S.$$

Where ΔG is Gibbs free energy change (kcal) which is the indication of the driving force of the reaction and is the degree of irreversibility if ΔG has a negative value. T and ΔS are the temperature (^{o}K) and entropy change (kcal/mol/^{o}K) respectively. However, this can be an over-simplified view of the energy changes during a reaction and a reaction should be considered not only thermodynamically but jointly with respect to the mechanisms by which bond linkages change.

3.2 Typical Process Requirements

For device fabrication the silicon to be etched can be that of a wafer (single crystal) or a layer of polysilicon used for MOS gate arrays. Polysilicon is usually deposited on a SiO_2 layer, then masked and finally etched. The etching normally requires selectivity to favour a fast etch rate for silicon with minimum loss of the SiO_2 substrate. This can be achieved by using a CF_4 + 3% O_2 plasma in a barrel reactor, with good selectivity between silicon and SiO_2, but the etching is isotropic. In a planar reactor, a CF_4 plasma etches less isotropically but also reduces the selectivity between silicon and SiO_2. Good selectivity between Si and SiO_2 and an anisotropic silicon etch can be achieved by using a chlorine plasma; but there is then usually poor selectivity between silicon and the masking photoresist masking layers.

In some device processes the SiO_2 layer (usually a dielectric) is required to be etched in favour of an underlying silicon layer which may have sensitive device regions. Using a CF_4 plasma in a planar or barrel reactor does not usually provide sufficient selectivity. However, by increasing the carbon to fluorine ratio in a plasma, i.e. by using C_2F_6 gas in a planar reactor, the selectivity begins to improve and by using C_3F_8 the SiO_2 layer can be etched in a planar reaction eight times faster than the silicon layer.

3.3 Fluorine Rich Plasmas

3.3.1 Oxygen Additions

Mogab et al.(43) first observed that when small amounts of oxygen were added to a CF_4 plasma the etch rate of silicon increased dramatically and the etch rate of SiO_2 only increased slightly, resulting in a selectivity of 10:1. This effect has also been observed when etching Si_3N_4, for which the etch rate increased rapidly (41), and has been monitored using optical emission spectrometry by Harshbarker et al.(42). Harshbarker et al. observed that the increase in etch rate of Si_3N_4 with the addition of O_2 to CF_4 plasma was accompanied by a large increase in fluorine.

Mogab et al.(43) used a planar reactor with a 13.5 MHz rf supply and a mass spectrometer analysing the gas in the effluent line. An optical emission spectrometer was used to observe the electronically excited $F^{\cdot}$ atoms $3\ S^2P \rightarrow 3P^2\ P^o$ with a sapphire window for viewing the plasma. Mogab et al. observed an increase in the optical emission intensity with the quantitative increase in the fluorine concentration $[F^{\cdot}]$ and correlated this with the increase in the silicon etch rate on addition of oxygen to the CF_4 plasma. Mogab et al. concluded that the F˙ atoms were the principal etching species. The increase in the silicon etch rate peaked at 12% oxygen addition and the fluorine concentration [F˙] increased to a peak at 23% oxygen addition as shown in Fig. 3.0.

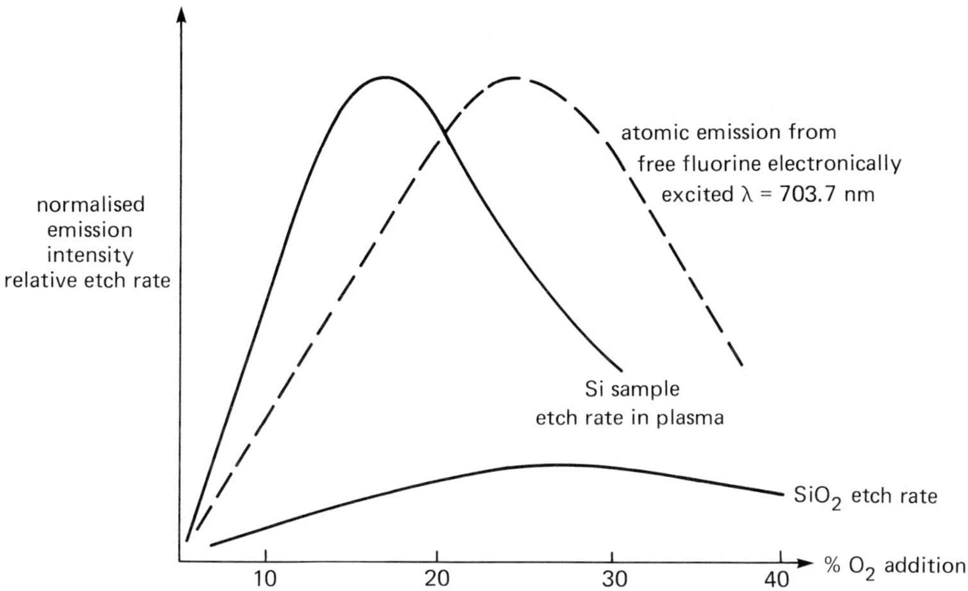

Fig. 3.0 Change in fluorine atom concentration and silicon etch rate with oxygen addition. (43)

Oxygen atoms do not react with CF_4 gas alone as the enthalpy of compound formation is insufficient. Oxygen can, however, react with dissociated CF_4 plasma products and can inhibit fluorine recombination reactions, typically by poisoning the surface to limit three-body recombination collisions:

$$\text{e.g. } F^{\cdot} + F^{\cdot} + M \text{ wall} \rightarrow F_2$$

$$CF^{\cdot} + F^{\cdot} + M \text{ wall} \rightarrow CF_4$$

An example of wall poisoning is the case when small quantities, 0.1%, of H_2O or N_2O gas were added to an oxygen plasma from which an increase in free radical concentration $[O^{\cdot}]$ was observed (44). This is due to surface poisoning by gas additions preventing three-body recombination events. The mass spectrometer used by Mogab et al. showed that silicon oxyfluoride was not formed in CF_4/O_2 plasmas but stable products of CO, CO_2 and COF_2 were observed. The role of oxygen is to scavenge the $CF_3^{\cdot}$ radicals thus preventing recombination with $F^{\cdot}$. Mogab et al. stated that the silicon etch rate was not strictly proportional to $[F^{\cdot}]$ because of the competition between $F^{\cdot}$ atoms and $O^{\cdot}$ atoms for the active sites on the silicon surface. Chemisorption of $O^{\cdot}$ on the silicon surface will reduce the etch rate whilst at the same time the presence of oxygen raises the $[F^{\cdot}]$ tenfold by preventing the following recombination reactions:

$$CF_3^{\cdot} + F^{\cdot} \rightarrow CF_4$$

$$CF_3^{\cdot} + CF_3^{\cdot} \rightarrow C_2F_6 \text{ (small unless } [CF_3^{\cdot}] \text{ is high)}$$

$${}^{*}CF_2^{:} + F_2 \rightarrow CF_4$$

Prevention of recombination of the fluorocarbon species by the scavenging action of O_2 and $O^{\cdot}$ is effective. Also the C = O bond has a high formation energy and stable products such as COF_2 can be formed as follows:

$$CF_3^{\cdot} + O_2 \rightarrow COF + OF^{*}$$

$$CF_3^{\cdot} + O^{\cdot} \rightarrow COF_2 + F^{\cdot}$$

OF^{*} is unstable and decomposes either by the process

$$2OF^{*} \rightarrow 2F^{\cdot} + O_2 \text{ or}$$

$$OF^{*} + CF_3^{\cdot} \rightarrow COF_2 + 2F^{\cdot}$$

---

*$CF_2^{:}$ is a diradical species formed in CF_4 plasmas with two unpaired electrons.

The addition of oxygen to a CF_4 plasma may also affect the ionization condition of the plasma. It has been observed that oxygen additions to pure He and N_2 plasmas cause a small change in the electron temperature (45). Small changes in electron temperature of a CF_4 plasma are not likely to affect the chemistry of the process. The peak in F$^\cdot$ concentration with the addition of 23% of oxygen observed in Mogab et al.'s experiment also happens to coincide with the peak etch rate of the slower etching silicon dioxide. From this it was concluded that $CF_3^\cdot$ radicals are not necessary for the reaction between SiO_2 and F$^\cdot$ atoms.

Horiike and Shibagki (46) devised an etching experiment using the apparatus shown in Fig. 3.1. They found that the etch rate of a silicon wafer, shown in Fig. 3.2 was sensitive to the distance from the CF_4 plasma (curve a). When O_2 was added to the CF_4 plasma the etch rate was higher at given distances from the discharge (curve b). When the silica tube was removed from the experimental apparatus and replaced with an alumina (Al_2O_3) tube the etch rate at given distances increased still further (curve c).

The increase in etch rate at given distances with oxygen additions is claimed to be due to the longer life times of the active fluorine species present. The life time increased slightly more by

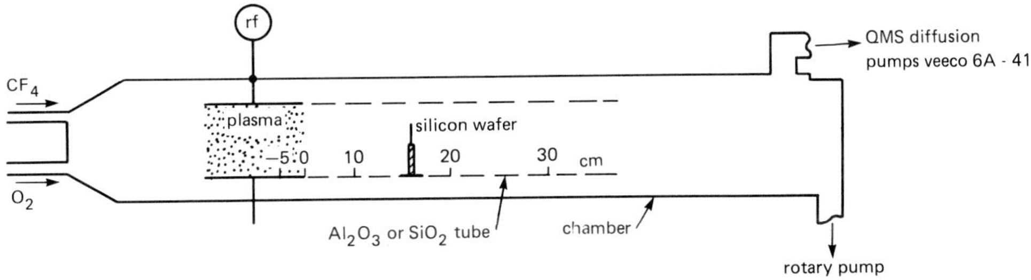

Fig. 3.1 Apparatus used by ref 46 for etching a silicon wafer at various distances from a CF_4 plasma source (46)

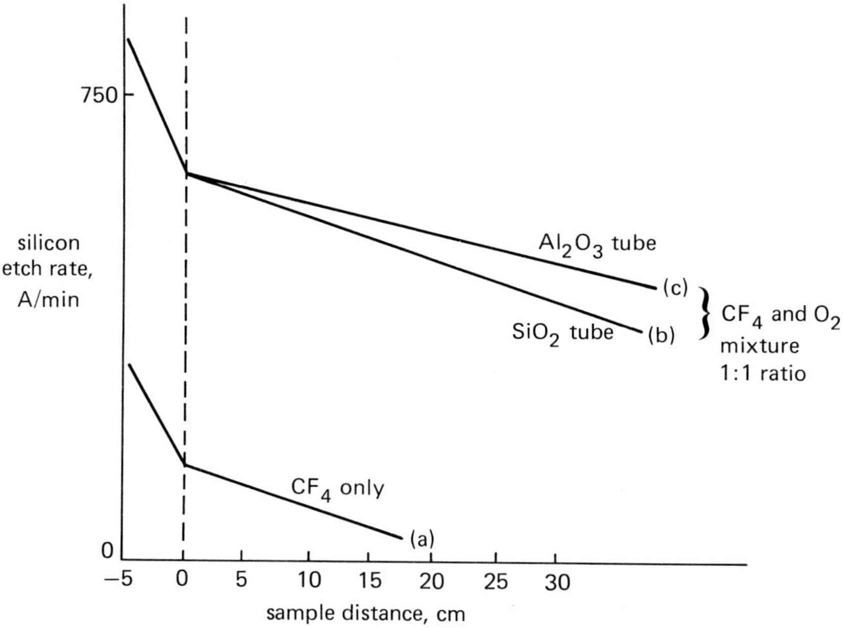

Fig. 3.2.
Silicon etch rate at set distances from the plasma, from ref 46.

replacing the silica tube with an Al_2O_3 tube since Al_2O_3 is more inert to the fluorine species. Horiike and Shibagki also found that etching at given distances was a maximum when the ratio of O_2 to CF_4 was 1:1 in the plasma, and concluded that under these conditions $COF^{\cdot}$ is an active species.

The generation of $F^{\cdot}$ at the surface is by the dissociation of $COF^{\cdot}$ according to a mechanism:-

$$4CF_4 + 3O_2 \xrightarrow[\text{plasma}]{} 6F_2 + F^{\cdot} + O^{\cdot} + COF^{\cdot} + COF_2 + CO + CO_2$$

Now

$$COF^{\cdot} \xrightarrow[\text{adsorption (surface)}]{\text{dissociative}} F^{\cdot} + CO\uparrow \qquad (A)$$

$$4F^{\cdot} + Si \rightarrow SiF_4$$
$$\text{surface}$$
$$F^{\cdot} + COF^{\cdot} \rightarrow COF_2 \qquad (B)$$

The drop in concentration of $F^{\cdot}$ in Mogab et al.'s experiment could be due to reaction - 'B', as large quantities of COF_2 were indicated by mass spectrometry data. The reaction 'A' proposed by Horiike and Shibagki requires further consideration and experimental evidence.

Flamm (47) has studied the oxidation of CF_4 and C_2F_6 using planar and tubular alumina reactors. He noted that when oxygen is added to a C_2F_6 discharge almost no increase in atomic fluorine was observed until a threshold concentration of 20% oxygen is reached.

The C_2F_6 discharge has a more complex chemistry and when C_2F_6 is ruptured by a collision with a plasma electron the formation of two $CF_3^{\cdot}$ radicals is energetically favoured, producing a low yield of $F^{\cdot}$

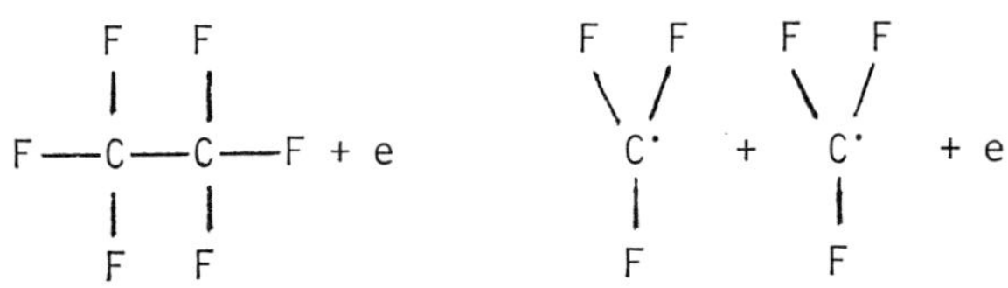

(Bond energies of C-F and C-C are 105 and 84 kcal per mole respectively.)

The high concentration of $CF_3^{\cdot}$ and the low $F^{\cdot}$ concentration in the C_2F_6 plasma prevents small additions of oxygen from generating $F^{\cdot}$. There is also likely to be a higher number of unsaturated compounds which will consume oxygen, e.g.

$$C_2F_4 + O_2 \rightarrow 2COF_2$$

Mogab (48) suggested that the thermodynamically favourable disproportion reaction to account for the $CF_2^{:}$ species is

$$CF_3^{\cdot} + CF_3^{\cdot} \rightarrow CF_2^{:} + CF_4 \ .$$

This has been observed by optical spectroscopy. The $CF_2^{:}$ is also sufficiently stable to be detected by mass spectroscopy (49). The $CF_2^{:}$ generated in a plasma from a high $CF_3^{\cdot}$ concentration will react with oxygen forming COF_2, e.g.

$$CF_2^{:} + O^{\cdot} \rightarrow COF_2 \ .$$

Heineke (50,8) using a planar reactor claims that in a CF_4 plasma a shift towards higher concentrations of $CF_3^{\cdot}$ would decrease the silicon etching rate which prefers $F^{\cdot}$. He assumes that $CF_3^{\cdot}$ takes an active part in the etching process by attacking the oxide and reducing the SiO_2 to the monoxide, which is converted to a siloxane and produces a volatile $(CF_3)_2O$ compound in the CF_4 discharge. However $(CF_3)_2O$ and siloxane compounds have not been observed in the mass spectra of a CF_4 discharge. In Mogab et al.'s experiment the SiO_2 etch rate increased with an increase in $F^{\cdot}$ concentration, from 100Å per min to a maximum of 350Å per min at 23% O_2 addition, after which the $F^{\cdot}$ concentration decreased, and so did the SiO_2 etch rate.

A plot of the $F^{\cdot}$ concentration (estimated by optical emission) versus SiO_2 etch rate is shown in Fig. 3.3. The intercept of the SiO_2 etch rate axis at zero $F^{\cdot}$ concentration shows a limited SiO_2 etch rate possibly due to the $CF_3^{\cdot}$ radicals or the ion bombardment contribution to the SiO_2 etching.

3.3.2 Other work

Beenakker et al. (51) investigated CF_4 plasmas with oxygen additions, using a mass spectrometer differentially pumped system using a microwave 2450 MHz plasma excitation. They concluded that oxygen atoms, and not molecules, were mainly involved in the reactions with the CF_4 plasma fragmented species and that CO was a primary plasma product which reacted with $F^{\cdot}$ and $O^{\cdot}$ to form CO_2 and COF_2. These experiments were

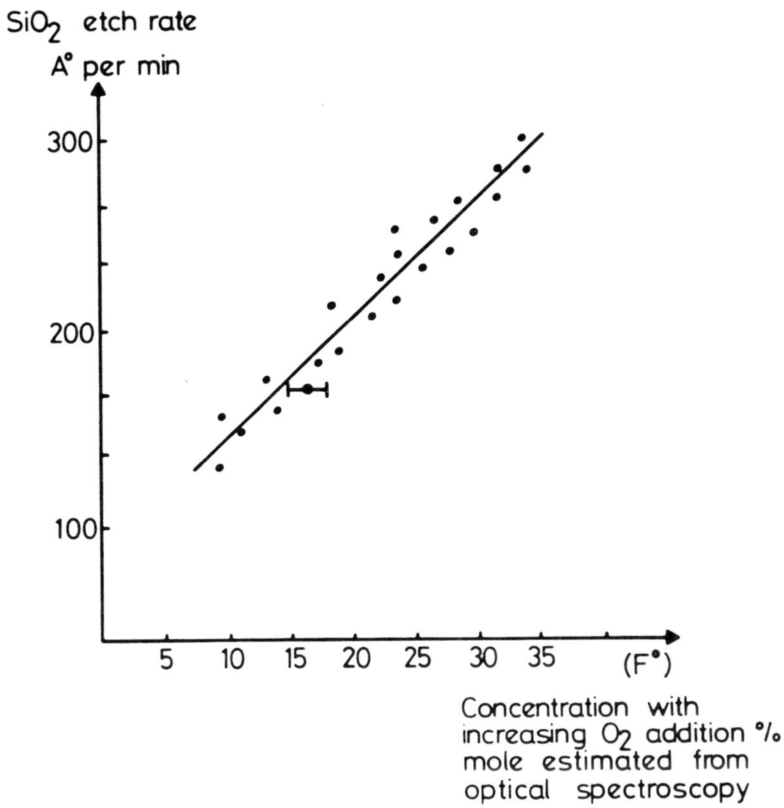

FIG 3.3

SiO_2 ETCH RATE VS F° CONCENTRATION ESTIMATED BY OPTICAL EMISSION (VARIED BY O_2 ADDITIONS TO A CF_4 PLASMA). (43)

carried out in after glow conditions and thus recombination events can change the levels of composition from the terminated plasma.

Flamm and Smolinsky (52), using a tubular fast flow rf reactor showed by mass spectrometry that CO forms in the early stage of the oxidation process and is subsequently oxidized to CO_2 and COF_2 at longer gas phase resident times. More recent work by Agostino et al.(53) has produced detailed emission spectra of CF_4/O_2 plasmas at 1 torr pressure and 50 watts rf power input. These results are in line with the general mechanisms suggested by Mogab et al.

Smolinsky and Flamm (52) observed that F_2 does not appear to be produced down-stream from the CF_4 discharge zone, probably because of preferred recombination with other atoms or molecules such as C_2F_4, $CF_3^{\cdot}$ $COF^{\cdot}$ and $CF_2^{:}$. Smolinsky and Flamm have also observed that the level of $CF_2^{:}$ in the effluent line diminishes in the presence of SiO_2. This result suggests that $CF_2^{:}$ could be an active species in the etching of SiO_2. Increasing the $CF_2^{:}$ concentration in plasmas can change the etching characteristics of silicon and silicon dioxide with the onset of polymerization problems; this will be discussed in the next section.

3.4 Fluorine Deficient Plasmas

We have seen that small amounts of oxygen added to a CF_4 plasma generate more free fluorine which improves the selective etching between silicon and SiO_2. Another development needed is an etching process for removing phosphorous doped SiO_2, i.e. phosphorous silica glass (PSG) down to the sensitive silicon areas. The PSG is used as a layer between silicon and the first level metal layer. If a CF_4 plasma is used to open up the contact windows in PSG then the sensitive device area is also etched because the selectivity between the SiO_2 and silicon is poor.

3.4.1 Hydrogen addition

Heinecke (54) added small amounts of hydrogen to a CF_4 plasma and found that the reverse of O_2 additions occurred because the silicon etch rate had decreased greatly whereas the SiO_2 had decreased by only a small amount. Thus, by adding hydrogen to the CF_4 plasma, it was possible to achieve better selectivity between SiO_2 and silicon. Heinecke also found that by increasing the carbon to fluorine ratio of the fluorocarbon gas used in the plasma the selectivity between SiO_2 and silicon increased as shown in table 3.

Table 3

Gas	C:F Ratio	SiO_2:Si Selectivity
CF_4	1:4	1:1
C_2F_6	1:3	3:1
C_3F_8	1:27	5:1
CHF_3	1:2	10:1

(using a planar type of reactor, ref. 54). The addition of hydrogen reduces the $F^{\cdot}$ concentration in the plasma by the formation of hydrogen fluoride (HF) and thus increases the carbon to fluorine ratio in the plasma.

Coburn and Kay (55), using a CF_4 plasma with H_2 additions, detected large quantities of HF in the effluent line. They also observed that C_2F_6 and C_2F_4 plasmas behaved similarly to a CF_4/H_2 plasma with respect to SiO_2:Si etching selectivity. Lehmann and Widmer (56) have postulated that the addition of H_2 to a CF_4 plasma forms HF which preferentially etches SiO_2 rather than Si and hence improves the SiO_2:Si selectivity. Although anhydrous HF is reported to be unreactive towards SiO_2 (57), it is possible that in the presence of radiation (from the plasma) the overall reaction:

$$SiO_2 + 4HF \rightarrow SiF_4 + 2H_2O$$

is promoted by radiation. The latter does not explain the increase in selectivity which occurs when the carbon to fluorine ratio increases without the addition of hydrogen, e.g. when C_2F_6 and C_3F_8 are used.

Coburn and Kay (55) added H_2O vapour to a CF_4 plasma in a planar reactor and observed that H_2O behaves as a mixture of H_2 and O_2 in the ratio of 2:1 respectively. They concluded that the behaviour of the discharge is not significantly influenced by the molecular structure of the injected gas but is determined primarily by the composition of the glow discharge. Coburn and Kay's work was for small additions of H_2O vapour at pressures below 100 mtorr which is not normally encountered in plasma etching.

3.4.2 Surface analysis and carbon

Coburn, Winters and Chang (59) investigated the etching behaviour of CF_3^+ bombardment (the most likely constituent ion in a CF_4 plasma) of silicon and silicon dioxide surfaces. They observed using XPS, Mass spec and AES analysis that resulting from the CF_3^+ impact, there was a carbon build-up on the silicon surface and this increased in thickness with increasing ion dose. A decrease in carbon thickness at higher CF_3^+ bombardment energies occurred possibly because of carbon sputtering, and the etch rate was independent of dose under these conditions.

At low ion bombardment energy the etch rate was observed to fall with an increase in dose of the CF_3^+ ions because the carbon film formed on the surface had increased in thickness. The carbon film was observed using AES and at high ion doses the film became very thick. At an ion energy of 500 eV the film formed was analysed using XPS data and C-F bond formation was observed. The C-F film coverage on the silicon surface was low for CF_3^+ ion energies above 500 eV for the same dose. Also above 500 eV XPS indicated fluorine atoms on the surface combined with silicon to form Si-F bonds. Below 500 eV and at the same dose the silicon surface was covered with a thick carbon film which prevented the fluorine atoms from directly bonding to the silicon surface and hence C-F bonding was observed. At energies above 500 eV the etch rates, measured in Coburn et al. experiments, were five to ten times faster than that from physical sputtering alone.

These experiments by Coburn et al. (59) could not be repeated with SiO_2 substrates because of the electrical charging of the dielectric when bombarding with ions. However they implanted oxygen into the silicon surface using O_2^+ ions at 4 KeV and when repeating the experiment found that carbon film formation was reduced. The etch rates were higher for a given energy and dose of CF_3^+ compared with a non-implanted silicon surface. The carbon film could have been removed by the oxygen in the silicon lattice, with ion bombardment supplying the necessary energy and allowing the fluorine to attack a relatively free silicon surface. Coburn and Kay (55) believe that O_2 additions to a CF_4 plasma increased the silicon etching by removing carbon from the silicon surface. Carbon film does not only prevent etching but also decreases the atomic fluorine concentration by recombination events.

An SiO_2 substrate etched in a high carbon to fluorine ratio plasma such as C_2F_6 would have less carbon on its surface than silicon. The high carbon content on the silicon will prevent etching. Providing the ion bombardment has sufficient energy to weaken and activate the Si-O bond on the surface, it is possible that the oxygen from the Si-O bond would react with the carbon film and reduce its thickness, so leaving more silicon in the SiO_2 available to react with fluorine species. Under the foregoing conditions the etch rate for SiO_2 should be faster than that of silicon and this selectivity has been confirmed by other workers (58), who also found that under certain conditions carbon film formation can extend to polymer deposition which terminates etching of both Si and SiO_2.

3.4.3 Polymer sputtering

Holland et al.(64) first observed both polymer formation and etching in an r.f. plasma in the vapour of a Fluorocarbon polyether. The plasma was contained in a silica vessel with external capacitively coupled electrodes. A polymer containing silicon oxide was formed on the vessel's floating potential regions, with etching occurring by F˙ and chemical sputtering at the zones where the electrodes produce a negative wall bias. Holland et al. have shown that polymer formation and etching are related to the ion impact processes.

Later Coburn and Winters (63) studied the effect on the silicon etch rate of hydrogen additions to a CF_4 plasma and observed a fine balance between etching and polymer formation at the wafer surface depending on the C:F ratio of the gas. The polymer can be sputtered by ion bombardment leaving the surface free to be attacked by species such as F˙ and $CF_3^{\cdot}$. The etching selectivity for silicon and SiO_2 with polymer formation is a balance, with removal rate on the SiO_2 surface being higher due to oxygen reacting with the carbonaceous polymer film. The effect of sputtering polymer away by increasing the negative bias on the sample increases the Si etch rate and has been reported in ref. 65 and is shown in fig. 3.4.

The electrode potentials developed in plasma reactors (14,15) can be as high as 100 to 1000 volts peak to peak. At a high plasma to wafer potential, i.e. high ion bombardment energies, some plasma systems will not experience polymer deposition until the ion bombardment energy is reduced.

Coburn and Winters (63), using a CF_4 plasma in a planar ion etcher, observed for silicon a decrease in etch rate with H_2 additions to a CF_4 plasma; at the same time the etch profile became more anisotropic with the H_2 additions. The silicon etch rate on the live r.f. electrode in H_2/CF_4 plasma increased as the negative bias to ground on the live r.f. electrode increased, coinciding with an increase in anisotropy of the etched silicon. This is due to the increase in the ion bombardment etching component as the negative bias to ground is

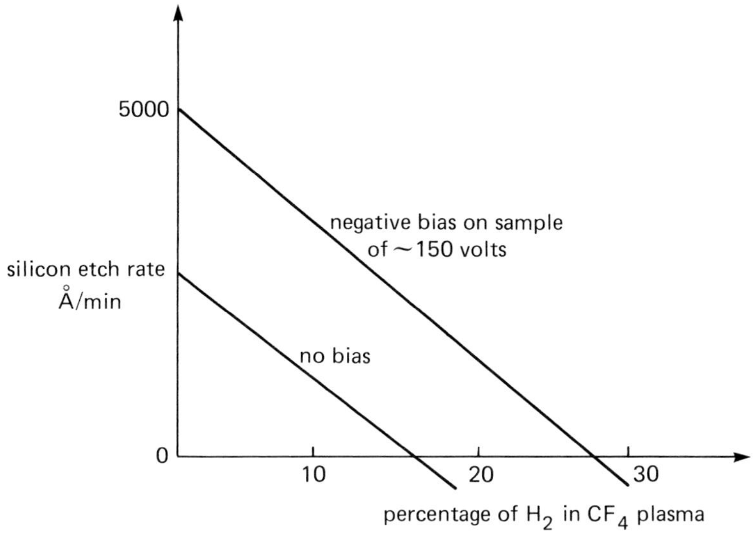

Fig. 3.4. Silicon etch rate at the live r.f. electrode surface vs % H_2 addition to a CF_4 plasma at zero bias and at a negative dc bias to ground of 150 volts. (65)

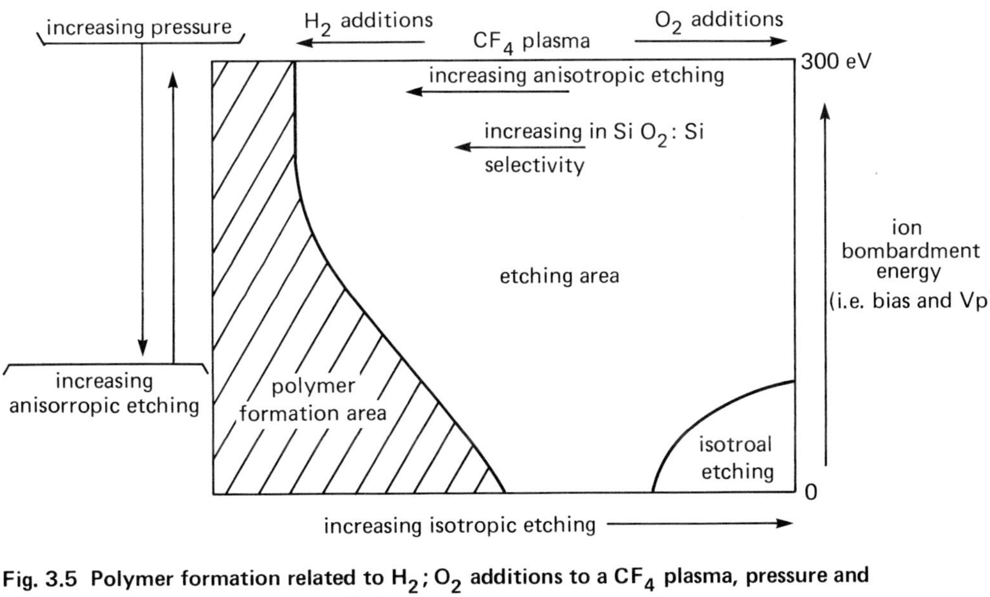

Fig. 3.5 Polymer formation related to H_2; O_2 additions to a CF_4 plasma, pressure and ion bombardment energy. (63)

increased. The decrease in silicon etch rate and increase in anisotropy with H_2 additions to a CF_4 plasma is due to the reduction in F˙ concentration by HF formation. Fig. 3.5 summarises Coburn and Winters' results for the conditions affecting etching and polymer formation.

Similar work was carried out by Matso (58) who observed that the amount of C_2H_4 needed to be added to a CF_4 plasma in a planar reactor before polymerization occurs increases with the increase in r.f. power input to the system. This is due to the increase in ion bombardment energies as the rf power increases which is sputtering the polymer deposit. Polymer formation is also sensitive to wafer loading effects as these can affect the C:F ratio of the plasma by wafer consumption of F˙ species. A large silicon area will more effectively consume the F˙ species and increase the carbon content of the plasma, thus inducing polymer formation.

Very thin (∿ 300Å) carbonaceous films have been observed on silicon wafers after etching in CF_4 plasmas and this can cause high contact resistance during subsequent processing. Such processing may involve removing the SiO_2 surface layer on the silicon substrate using a CF_4 plasma and then depositing sputtered aluminium films on the silicon surface. However, the etched silicon appears to have a thin carbonaceous film, itself giving a poor contact resistance (66). To reduce this, small amounts of oxygen can be added to reduce any film formed. However, CF_4/O_2 plasmas can only be used if selectivity is not a problem.

In previous discussions polymer or carbon film deposits, when present, give good selectivity between SiO_2 and Si, since the films preferentially deposit on the silicon surface and can inhibit the etching of any sensitive device areas. It is believed that such a film formation must be rapid and uniform otherwise roughing of the silicon surface occurs (66). After etching a SiO_2 layer down to a silicon substrate using a CHF_3 plasma in a planar reactor, there remains on the exposed silicon surface a thick polymer or carbon deposit which can be removed by "ashing" in an oxygen plasma. Alternatively, the deposits can be burnt off in an oxygen atmosphere at 600 ℃. Some workers believe that these treatments still leave a thin film on the surface which is not removed by HF solutions and may be silicon carbide formation (67).

CF_4 plasmas generate some C_2F_6 (47) which, at a high concentration, can produce polymer films. However, when O_2 is introduced into the plasma, C_2F_6 is not observed; instead, formation of COF_2, CO and CO_2 is evident from mass spectrometer data. Small air leaks in some systems may prevent the onset of polymer deposition. Also small additions of either He or Ar can alter the ionization conditions (45) in the plasmas which may change the onset of polymerization because these gases have high ionization potentials and form metastables.

3.4.4 Carbon free gases

The addition of O_2 to non-carbon containing gases where a carbon deposit would not occur can also increase the etch of silicon, e.g. SiF_4/O_2 mixtures, see ref. 38.

$$SiF_4 + O^\cdot \rightarrow SiOF_2 + 2F^\cdot$$
$$\rightarrow \text{Silicon Oxyfluoride (Volatile)}$$

and $2F^\cdot$ produced in plasma will be available to etch silicon. This process gives a good selectivity between $Si:SiO_2$ but the etch rates are very small. The addition of O_2 to a sulphur hexafluoride plasma (SF_6), ref. (60), increased the silicon etch rate.

$$Si + SF_6 + O_2 \rightarrow SO_2 + SiF_4 + 2F^\cdot$$

The increase in silicon etch rate with O_2 additions to a SF_6

plasma was accompanied by a small increase in SiO_2 etch rate. The maximum F$^\cdot$ concentration ([F$^\cdot$]) generated by a SF_6/O_2 mixture occurred at ∿ 40% O_2 addition to the SF_6 which also coincided with the optimum selectivity for Si:SiO_2 of 70:1. The [F$^\cdot$] maximum in the CF_4/O_2 plasmas was at 12% O_2 addition compared with SF_6/O_2 plasmas which was at 40% O_2 addition, and the amount of F$^\cdot$ production at the maximum for the SF_6/O_2 plasma was larger than the maximum for the CF_4/O_2 plasma. The greater amount of F$^\cdot$ produced in a SF_6 plasma is due to the more extensive dissociative process. Eisele (61) also found that the etch rate of silicon was raised when O_2 was added to a SF_6 plasma, but certain experimental points differ from similar work by Flamm and Mogab (62) who, using mass spectroscopy, observed SO, SOF and SOF_2 products.

3.4.5 Electrode effects

Increasing the electrode separation distance may increase the plasma volume in contact with the vessel walls and if these walls are at ground potential then the ground electrode area will increase with respect to the live rf electrode area. Thus the negative dc bias formed on the live r.f. electrode will increase. The polymer films deposited during plasma etching will be reduced when the electrodes are separated, for two reasons. Firstly, polymer deposition will be dispersed over a larger chamber area, and secondly, at the live electrode polymer films will be subjected to an increase in ion bombardment energy (66). The ground electrode may also be subjected to a smaller increase in ion bombardment energy, providing the dc negative bias on the rf electrode is not too large, since there will be an increase in applied potential. The increase in applied potential when the electrode separation distance is increased is due to the higher plasma recombination events between the electrodes, thus increasing the impedance, and hence a higher potential is required to maintain the same charge flow.

The ionisation conditions of plasmas can be changed by changing the electrode material. This was observed by M. Gill (132) where stainless steel electrodes were replaced by silicon electrodes of equal area in a planar reactor using an Ar plasma at 100 mtorr pressure. He observed a voltage increase to maintain the same current density, as shown in fig. 3.6 which appeared to be related to the secondary emission coefficient (γ_i) for the electrode material.

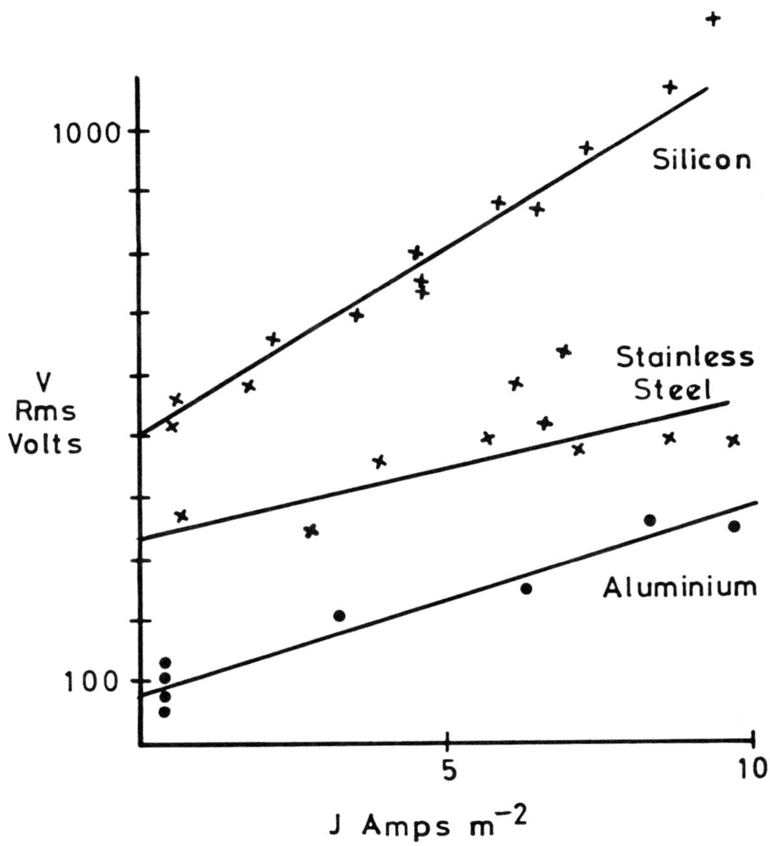

FIG 3.6

VOLTAGE/CURRENT DENSITY PLOTS FOR DIFFERENT ELECTRODE MATERIALS USING Ar PLASMA FROM REF 132

In low frequency plasmas, ionizations are sustained by secondary electrons which are generated by ion impact on the electrode surface and accelerated out into the plasma by the dark space sheath. When electrodes in a planar reactor are replaced by material with a lower γ_i, then to maintain the same ion density the voltage must increase across the dark space. The smaller number of secondary electrons will then be accelerated with more energy to maintain the same ion density.

The author (130) considers the possibility that secondary electrons leave the aluminium electrode surface, are accelerated by the ion field and collide with molecules to produce free radicals and ions within the dark space (Multiplication). This creates a high concentration of energetic free radicals, increasing the likelihood of polymer deposition. This, however, would not explain polymers forming on surfaces other than the electrodes in the system, since some surfaces may only have a small dark space sheath potential and width, hence the multiplication process may not occur so readily.

Electrodes made of aluminium appear to increase the likelihood of polymerization compared with that for stainless steel electrodes. This is believed to be caused by aluminium halides formed in the plasma, acting as a Ziger Catalyst increasing the molecular weight of the free radical species (133). Coating the chamber wall electrodes with a polymer film can be achieved by using gases of high carbon to fluorine ratio such as C_4F_8. Coating by this method before loading and etching the wafer will reduce any metal electrode effects which may cause polymer formation on the samples to be etched (133).

Good selectivity between SiO_2 and Si can also be achieved even when using a CF_4 plasma, if it is pre-conditioned with a C_4F_8 plasma because the unsaturated polymer (CF_x) can remove fluorine, hence reducing the $F^{\cdot}$ concentration, and increase the effective C:F ratio (68).

3.4.6 RIBE and sputtering

Mayer et al.(69) used RIBE and bombarded a SiO_2 sample with Ar^+ ions with the sample in a background gas of CF_4*. They observed that, at high beam energies, the etch rate was at a maximum when the incident beam was at 45° to the sample surface plane. At lower ion energies the maximum etch rate occurred at 0°; so they concluded that at high energies momentum transfer played the major etching role whereas at low beam energies a chemisorption process occurred, working in combination with the incident ions, allowing reaction to proceed. This was also shown by the etch rate of SiO_2, at fixed Ar^+ ion bombardment energy, rising as the CF_4 pressure was increased, shown in fig. 3.7 .

The fluorocarbon CF_4 has a low sticking coefficient for SiO_2 (70), but $CF_2^{:}$ released in a plasma could have a high sticking coefficient to SiO_2 and in large amounts could be responsible for polymer formation (71). The increase in etch rate with CF_4 pressure in fig. 3.7 could be due to the increase of adsorption of CF_4 or from the increase in $CF_2^{:}$ production. It would be of interest to know what the effect on the adsorption processes would have been if the sample temperature had been varied in the experiment of Mayer et al.(69). However, fig. 3.7 indicates that impacting Ar^+ ions were inducing chemical etching of SiO_2 probably by the fragmentation of adsorbed CF_x species on the surface. If the CF_4 partial gas pressure was replaced by Ar gas the etch rate was considerably reduced and the etching stopped if the Ar^+ bombardment stopped with or without CF_4 background pressure.

---

*CF_4 gas is chemically inert to SiO_2 and Si surfaces but may be weakly adsorbed.

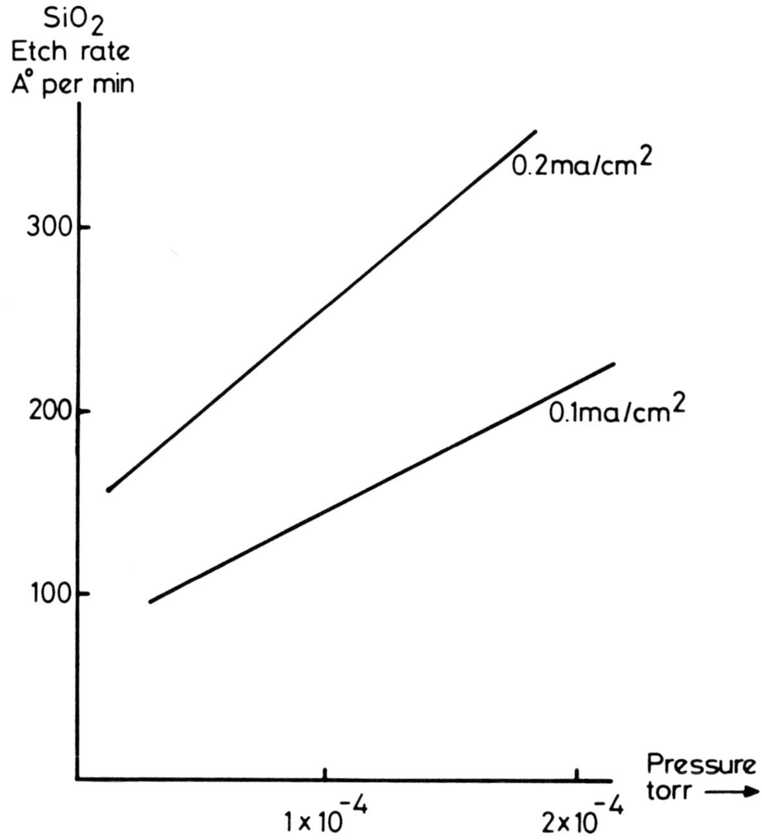

FIG 3.7

SiO_2 ETCH RATE VS CF_4 BACKGROUND PRESSURE AT Ar^+ CURRENT DENSITIES OF 0.1 AND 0.2 MA/CM^{-2} (69).

3.5 Ion V Free Radical in Plasma Etching

Coburn and Kay (55) and other workers (64,56) proposed that, in the etching of silicon in a CF_4 plasma, there was a co-operative interaction between ion bombardment and free radicals. Ion bombardment may remove any carbonaceous deposit and damage the surface, thus increasing the probability for a free radical to react with the exposed and weakened silicon bonds. Winters (70) observed that neutral $CF_3^{\cdot}$ radicals (generated from a CF_4 plasma) have a reaction probability of only 0.25 whereas CF_3^+ has a reaction probability of 1 on a clean silicon surface. His results show that the ions strongly react with silicon and thus influence the etch rate considerably.

3.5.1 Chemisorption

In section 3.1 the bond energies of Si-F (130 kcal per mole) reveal a thermodynamically favourable reaction for:

$$2F_2 + Si \rightarrow SiF_4.$$

However this reaction proceeds slowly at room temperature with a reaction coefficient of only $\sim 10^{-5}$ (28), but the reaction rate does increase with a rise in pressure and temperature.

Coburn et al.(63) have observed that XeF_2 gas reacts with silicon four times faster under the same conditions (pressure and temperature) than F_2 gas. This is due to a higher rate of absorption of atomic fluorine species when XeF_2 is used. XeF_2 dissociatively chemisorbs on silicon, with Xe being desorbed and leaving larger $F^\cdot$ concentrations available for a reaction. Flamm et al.(72) have shown that reactions between fluorine and silicon are reaction rate limited by the chemisorption process. Chemisorbed groups may be ultimately released by the reactions of impact ions weakening the Si-Si bond and allowing the reacting species to be directly chemisorbed which would otherwise be a slow process:-

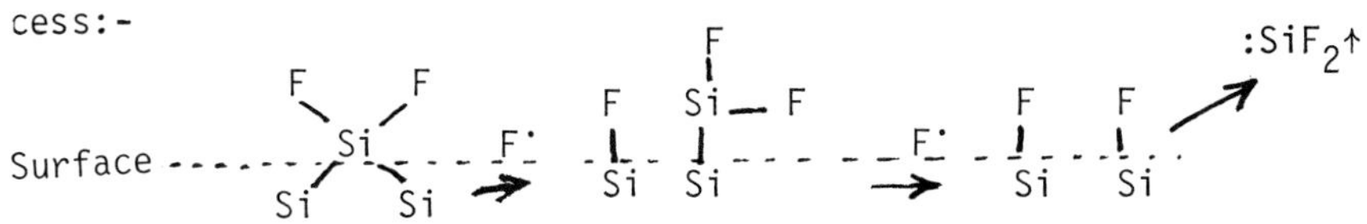

Flamm et al. state that the rate determining step is in the surface formation of $:SiF_2$ which is released by atom impact onto the silicon surface. Flamm et al. have estimated in detail the complex kinetic mechanisms for the interaction between $F^\cdot$ atoms and a solid surface of Si and SiO_2. They state that the F· atom - solid reaction alone can account for the etch rates in plasmas at a few tenths of a torr pressure range. They claim ion or electron bombardment does not play an essential role in etching Si or SiO_2 as this is consistent with the isotropic nature of the F· type etchants. However in Flamm et al.experiments the etch rates for Si and SiO_2 are very small and the isotropic free radical etching of SiO_2 is not consistent with other workers (67,55).

Coburn et al. (63) have observed that SiO_2 does not react with XeF_2 gas or F_2 alone at ambient temperatures but when a SiO_2 sample is exposed to a beam of ions or electrons, then etching proceeds. Their results are displayed in fig. 3.8. Although SiO_2 did not etch on exposure to XeF_2 in the absence of an electron bombardment, Auger spectroscopy showed a layer of absorbed fluorine present. A similar experiment by Flamm et al. revealed some Si-F bond formation by infra-red surface spectra analysis. The results by Flamm et al. indicate that dissociative chemisorption of XeF_2 had occurred on the SiO_2. Etching occurs on this layer of SiO_2 with an absorbed or chemisorbed F· layer when subjected to electron or ion bombardment. This irradiation enhances the etching probably by bond damage of the Si-O whilst not appearing to sputter away any absorbed F· layer formed from the XeF_2.

Etching SiO_2 in a CF_4 plasma using a barrel reactor gives a very low etch rate because of the absence of ion impact. Some etching will occur from the F· radicals and $CF_3^{\cdot}$ radicals produced. However silicon reacts very much faster with F· than $CF_3^{\cdot}$ radicals because $CF_3^{\cdot}$ radicals have a very low reaction probability with Si of $\sim 10^{-5}$. Thus selectivities greater than 40:1 for Si to SiO_2 in a barrel reactor may be achieved. Etching Si and SiO_2 samples in a CF_4 plasma using a planar reactor produces ions and free radicals both of which are capable of etching the samples.

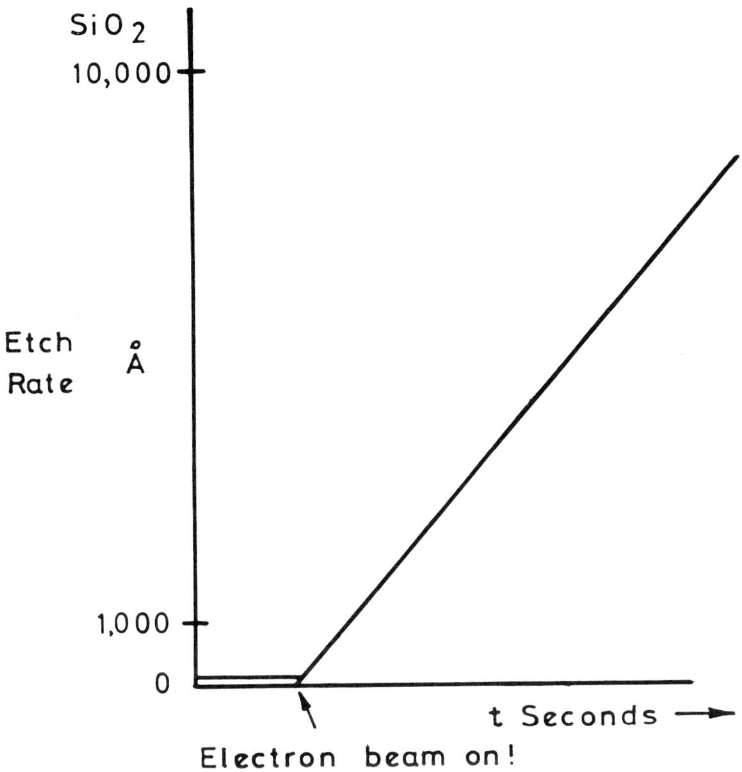

FIG 3.8

SiO_2 ETCHED (Å) IN XeF_2 VS TIME (SECONDS) (63)

3.5.2 Ion impact

The main component of the electron fragmentation of CF_4 will be CF_3^+ because of the stable planar structure of the ion. Thus CF_4^+ will not exist because of the high CF_3^+ stability. Coburn et al. (63) extracted and analysed ions from a CF_4 plasma and observed the largest peak for CF_3^+ with some ionized dimer products, $C_2F_5^+$ peaks which diminished when SiO_2 was etched in the CF_4 discharge.

In the presence of a positive ion sheath at the work surface, the ions will obtain kinetic energy and bombard the surface substantially normal to the surface (assuming ions suffer negligible collisions in the dark space sheath). This would result, as shown in Fig. 3.9(a), in anisotropic etching. The ion induces reactions on impact, providing the reaction is fast and any fragmented free radicals produced are consumed by the reaction in the bombardment vicinity. Free radicals produced in plasmas are not influenced by the dark space field and, in sufficient concentration, can etch laterally, resulting in isotropic etching with characteristic undercutting of the mask area as shown in fig. 3.9(b).

From Fig. 3.9(b) an isotropic etch, such as in substrate etching with certain wet chemicals, can be defined by the relation A = 1 when x = d, $\left(A = \frac{d}{x}\right)$ and a totally anisotropic etch is defined as $A = \infty$, that is, there is no undercut or line width loss (x = 0). However, in plasma etching the etched characteristics are not pure isotropic or anisotropic profiles. For example, in a barrel reactor using a CF_4 plasma, silicon is etched to an almost isotropic profile, (A just a little above unity).

When a process shows a value of A larger than 1 some workers call this an anisotropic process, since there is a difference between the vertical and horizontal components. Other workers may call this an almost isotropic etching process. The small increase in vertical etch component compared with the horizontal etch component of silicon in a barrel reactor is possibly due to UV radiation enhanced vertical etching or, a difference between the active species adsorption coefficients for the silicon lattice planes during etching. Almost pure anisotropic etching can be obtained when etching proceeds by <u>ion bombardment inducing surface reactions</u> without any independent free radical contribution.

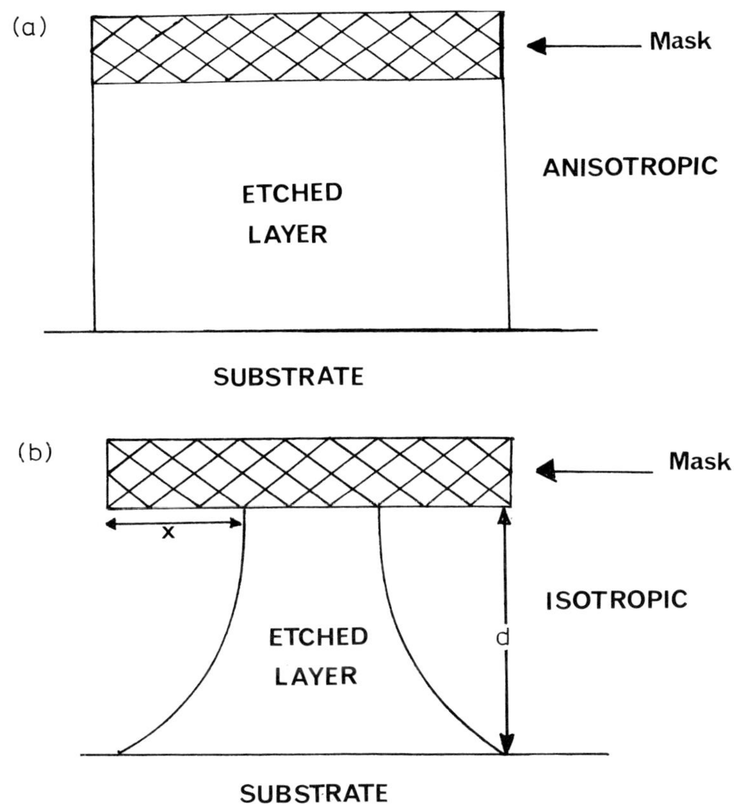

Fig 3.9

Examples of Isotropic and Anisotropic etched layer

A process such as the etching of polysilicon layers using chlorine gas in a planar reactor can achieve very high values of A, (A > 1000). The problem is in the measurement of small values of x and making these measurements assuming the mask dimensions have not changed. For example, some processes appear to have a very high A value, but the resist mask layer, if bevelled, will recede during etching giving an apparent coincident edge between the mask and etched layer, making it difficult to obtain x values. Ion bombardment can also during etching sputter away from the substrate surface any polymer or resist deposits which may redeposit onto the side walls of the etched profile thus reducing the free radical lateral etching component (73).

With any plasma process the type of etch profile will depend on the relative contributions from free radicals and ion impact. If etching was totally ion impact dependent then a change in free radical concentration should not affect the etching. In CF_4 plasma etching of silicon in a planar reactor, the etching is both free radical and ion impact enhanced. Addition of O_2 to a CF_4 plasma increases the free radical $F^{\cdot}$ concentration, and will make the etching more isotropic. Addition of H_2 to a CF_4 plasma will reduce the $F^{\cdot}$ concentration by H-F formation thus making etching more anisotropic. Etching silicon in CF_4/H_2, or gases of high carbon to fluorine ratio will be slow due to polymer deposits being removed solely by ion bombardment. Etching SiO_2 in a CF_4 plasma using a planar reactor will proceed with a high degree of anisotropy since etching is enhanced by surface ion bombardment and the free radical reaction coefficient is low.

Lehman and Widmer (74), using a CF_4 plasma at 0.1 torr pressure, etched a SiO_2 sample placed on the live electrode of a RIE with a dc negative bias of 200 volts. The SiO_2 samples were etched anisotropically both at the centre and at the edge of the work holder. At the edge of the work holder the electric field in the dark space is not normal to the work holder (electrode) surface and this was reproduced in the silica etching profile as shown in fig. 3.10. The ions tend to follow the electric field direction and bombard the surface, but at the edge of the work holder the ions bombard the surface at an angle causing a distorted etched profile in the SiO_2 layer. Lehman and Widmer concluded that

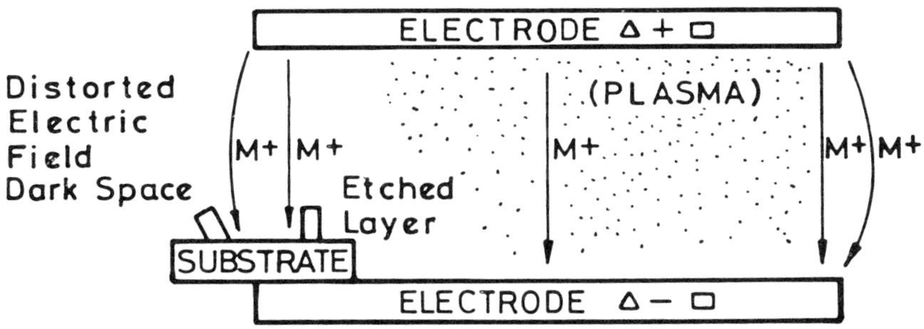

FIG 3.10

DIAGRAM SHOWING THE DISTORTED ELECTRIC FIELD AT THE EDGE OF THE WORK HOLDER WHICH IS THE LIVE RF ELECTRODE IN REF 74

ions play the major role in etching SiO_2 in a CF_4 plasma and other workers (75), with similar experiments, have reached the same conclusion.

Vasile (76) using an He/F_2 plasma instead of carbon containing halogen gases, found that the Si etch rate was faster on the live electrode than on the ground electrode of an rf planar reactor. The higher etch rate at the live electrode arose from the greater energy conveyed by the ion bombardment which also resulted in an increase in anisotropy for silicon etching, as shown in fig. 3.11.

S. Matsuo (58), using a RIE with a C_2F_6 plasma at 0.02 torr pressure, observed uniform etching rate for a SiO_2 wafer. When a Si wafer was placed onto a spacer 1.5 mm high and allowed to overhang by 5 mm, and then placed onto an SiO_2 surface then the observed etching of SiO_2 was non-uniform. The etching stopped in the SiO_2 layer where it was masked by the Si wafer, as shown in fig. 3.12. Matsuo visually estimated the dark space width in his experiment to be 1-2 cm between 0.2 to 0.001 torr. The SiO_2 area directly exposed to the plasma was

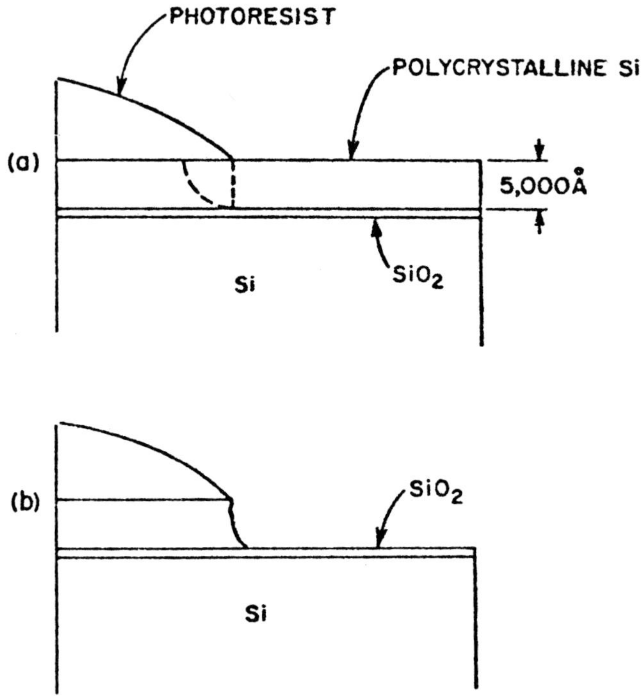

Fig 3.11

From ref 76. Tracings of the edge profile of SEM micrographs of a Si/SiO /polysilicon/photoresist substrate. (a) Magnification 20,000X, dotted lines at the photoresist/poly-Si interface show the edge profiles expected for isotropic and anisotropic etching (b) Edge profile obtained from etching the specimen in the 3% F -He plasma on the rf electrode

etched as shown in fig. 3.12 from which Matsuo concluded that ion enhanced etching was the main mechanism for etching. Matsuo observed a constant etch rate independent of sample temperature in a C_2F_6 plasma when etching Si and SiO_2, as shown in fig. 3.13.

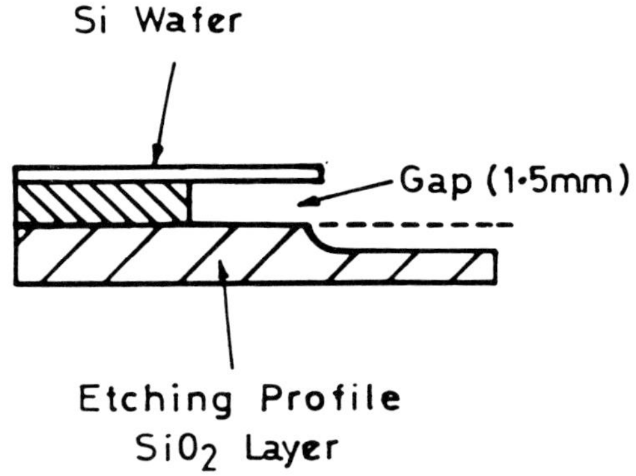

FIG 3.12

ETCHING PROFILE OF SiO_2 PARTIALLY MASKED WITH A GAP OF 1·5mm. ETCHING TABLE, FUSED QUARTZ, CF_4 PRESSURE 3×10^{-2} TORR FROM REF 58

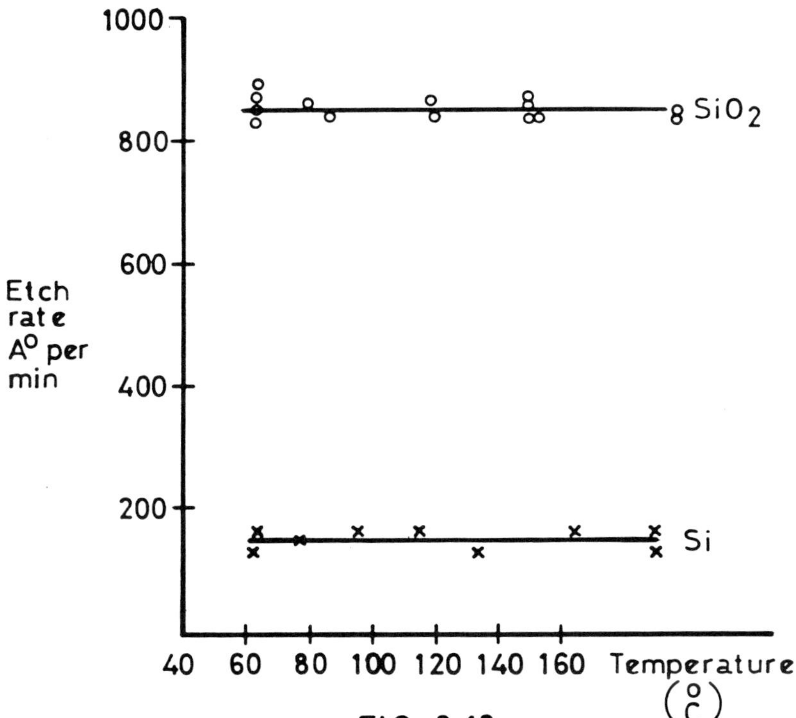

FIG 3.13

ETCHING RATE OF Si AND SiO_2 IN A C_2F_6 PLASMA (25cc/min, POWER 200WATTS) AT VARIOUS SAMPLE TEMPERATURES (58).

The constant etch rate observed by Matsuo for silicon is probably due to the low free radical concentration in a RIE system at the low pressure used. Since Matsuo used C_2F_6 gas, the polymer build up on the surface would have prevented a reaction between $CF_3^{\cdot}$ and $F^{\cdot}$ with silicon. Reactions would only occur when ions bombarded the surface breaking up the polymer film and inducing chemical reactions which would not have been affected by changes of sample temperature within the small range used by Matsuo.

3.5.3 Method of detecting undercut

A simple method used to detect undercut was to mask narrow channels in the sample and then place the sample upside down in the plasma. Thus only free radicals entering along the side of the slice can etch the sample as shown in fig. 3.14. This system, developed by M. Gill (77), will only work on surfaces which do not require ion bombardment to sputter away any native oxide films and hence will not work for aluminium or titanium etching.

3.5.4 He additions to CF_4 plasma

Schwartz (78) has found that additions of helium to CF_4 plasmas increases the anisotropic component of etching. This may be due to the small cross sectional area and the high ionization potential of helium which increases the plasma potential (V_p) of the system so maintaining a constant ion density. An increase in V_p will increase the sheath potential and thus increase the ion bombardment potential.

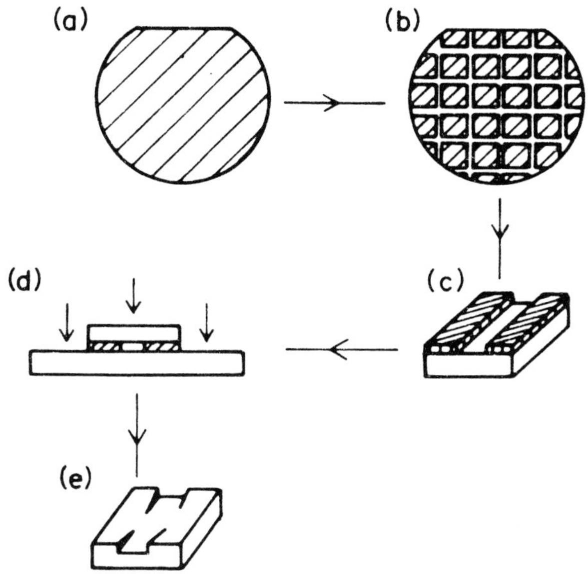

Fig 3. 14

Schematic processing sequence; (a) oxidise Si slice; (b) Wet etch channels; (c) cleave sample piece; (d) Plasma etch with sample face down on Si; (e) Remove oxide in HF acid and measure step at edge (77).

3.5.5 Effects of excitation frequency and pressure using a CF_4/O_2 plasma

Parry and Roddle (79) etched silicon using a CF_4/O_2 plasma at 42 mtorr on the ground and live rf electrodes of a planar reactor and observed a decrease in the anisotropic etching component (A) as the rf excitation frequency was increased. This was less noticeable on the live rf electrode, shown in fig. 3.15. They also observed a decrease in the anisotropic component for silicon etching as the pressure was increased, the frequency being fixed at 8 kHz.

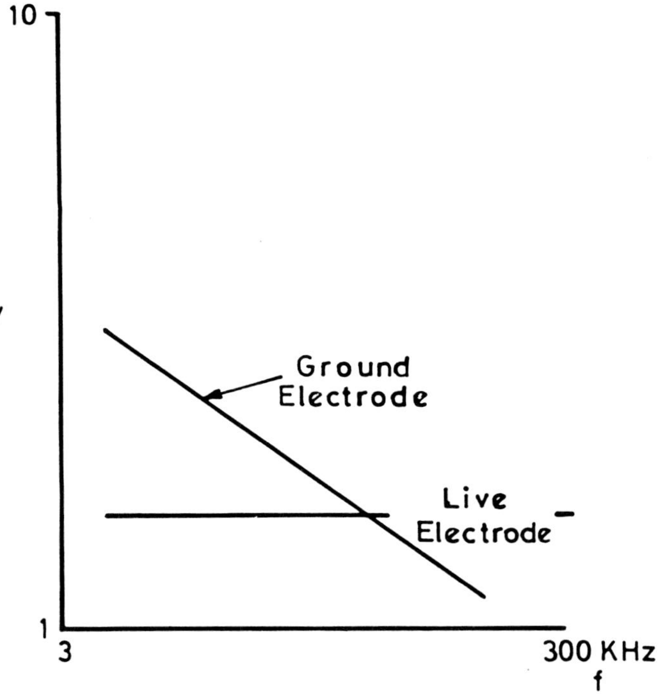

FIG 3.15

(A) VS EXCITATION FREQUENCY AT 42 MTORR USING CF_4/O_2 (8% O_2) (79)

The decrease in anisotropic etching of Si at the ground electrode as the rf excitation frequency is increased is probably due to a decrease in V_p at higher frequencies which reduces the ion bombardment energy. This reduction in V_p at the live electrode will not be so noticeable since there is probably an increase in dc negative bias potential on the live electrode as the frequency increases which will in turn increase the ion bombardment energy at the live electrode (132). The decrease in anisotropy as the pressure increases is probably due to the decrease in V_p accompanied by an increase in free radical generation.

3.6 Carbon Containing Mixed Halogens (CF_xXn) in Plasmas

Fluorine is the most electronegative halogen with the strongest bond to carbon: C-F> C-Cl> C-Br > C-I. (This order may change if the carbon is aromatic since there may be the possibility of π-d bond overlap with Cl, Br and I.) In the plasma inelastic electron collisions with CF_xX_n molecules will excite, ionize and rupture bonds. The carbon atom CF_xX_n will tend to hold onto the fluorine and release X_n during inelastic electron collisions.

$$CF_xX_n \xrightarrow[\text{bond rupture}]{e} CF_x^{\cdot} + X_n^{\cdot}$$

$$CF_xX_n \xrightarrow[\text{and ionization}]{e} CF_x^{+} + X_n^{\cdot}$$

The $CF_x^{\cdot}$ radical may polymerize, and this will reduce the effect of fluorine chemistry in the plasma and the X_n chemistry will tend to dominate, depending on its level of concentration. Matsuo (80) observed that additions of iodine vapour (20%) to a CF_4 plasma at 0.02 torr in a RIE at 100 watts gave no undercutting when etching silicon and increased the silicon etch rate. The SiO_2 etch rate at the same conditions decreased giving selectivities between Si:SiO_2 of up to 5:1. Matsuo suggests that iodine adsorbs on the SiO_2 surface and acts as a buffer during etching.

The rapid aniostropic etching of silicon is a mystery. However, iodine has a low ionization energy and a large cross sectional area thus there may be appreciable I^+ bombardment which will increase the anisotropic etch rate. There may also be selective plane adsorption of I_2 on Si thus inhibiting free radical reactions in the lateral directions, thus reducing undercut.

Matsuo, in the same reactor, used a $CBrF_3$ plasma at 0.03 torr pressure, 100 watts rf power. He etched silicon, silicon nitride and molybdenum aniostropically and obtained selectivities for $Si:SiO_2$ of up to 8:1.

Mogab and Shankoff (24) used $CBrF_3$ for etching titanium at 0.35 torr (200 watts) in a planar reactor and found that it was unstable unless 25% helium was added. In addition 12% O_2 was incorporated to help increase the atomic bromine [Br·] concentration in much the same way that O_2 is added to CF_4 plasmas. The rise in [Br·] with O_2 additions increased the etch rate of titanium and is believed to be the major active species in etching titanium.

Flamm et al. (81) observed in a $CBrF_3$ plasma a colour variation across their planar reactor indicating inhomogeneity. At 20 ml per min flow rate, low power (40 watts), etching of silicon was achieved. Polymer was deposited around the chamber and a murky green colour was formed near the gas feed. At a lower flow rate, 15 ml per min, high power (100 watts), there was more polymer formation accompanied by a colour change near the gas feed. The polymer films deposited were analysed by XPS (X-ray Photon Spectroscopy) and RBS (Rutherford Backscattering Spectroscopy). From these techniques, Br was not observed in the film itself, only at the silicon/polymer interface, on samples which had been etched at 20 ml/mm, 0.5 torr pressure, 40 watts rf power. Optical emission data from Flamm et al. showed no measurable concentrations of F_2 or F·, only Br_2, Br·, $CF_2^{:}$ and $CF_3^{\cdot}$ radicals. The high concentrations of $CF_3^{\cdot}$ and $CF_2^{:}$ are responsible for polymer deposition in the work of Flamm et al.

Beenaker et al. (82) have compared the mass spectrometer data with a CF_4 plasma and a $CBrF_3$ plasma and have concluded that the CF_4 plasma produces CF_2^+ ions. There was little evidence of CF_2^+ ions generated in the $CBrF_3$ plasma which is probably because any CF_2^+ generated can be reacted with $CBrF_3$ to release F· and Br· (83).

e.g. $$CF_2^+ + CBrF_3 \rightarrow C_2F_4^+ + F^\cdot + Br^\cdot$$

$$CF_2^+ + CBrF_3 \rightarrow C_2F_5^+ + Br^\cdot$$

CF_2^+ acts as a scavenger ion, but is unlikely to react with CF_4 in the plasma since:-

$$CF_2^+ + CF_4 \rightarrow C_2F_4 + F_2$$

is an endothermic reaction. Large quantities of C_2F_4 were found in the $CBrF_3$ plasma, accompanied by some CF_2Br^+, probably produced by the reaction with a CF_3^+ ion:-

$$CF_3^+ + CF_3Br \rightarrow CF_2Br^+ + CF_4$$

$CBrF_3$ plasmas behave very differently from CF_4 plasmas. Many workers have observed anisotropic etching of silicon by $CBrF_3$ plasmas at pressures between 0.2 and 0.5 torr (67). This is due to the C-Br bond being weaker than the C-F bond and more prone to cleavage in the plasma.

$$CBrF_3 \xrightarrow[\text{collision}]{e^\cdot} CF_3^+ + Br^\cdot + 2e^\cdot$$

The $Br^\cdot$ atom has a large electron cross sectional area and can easily capture low energy electrons to become Br^-; this can lead to $CF_3^+ + Br^-$ recombination giving the plasma unstable properties. The large Br^- ion may also collide with F_2 or $F^\cdot$ at high pressures and, since fluorine is more electronegative, it will remove the charge from Br to form F^-.

$$Br^- + F^\cdot \rightarrow F^- + Br^\cdot.$$

F^-, once formed, will not play an important role in any etching as it will be held by the positive plasma, not being able to overcome the plasma potential. This reaction, forming F^- by collisions with Br, will reduce the $F^\cdot$ concentration, as will other recombination reactions such as F· reacting with $CF_2^{\cdot}$ and $CF_3^{\cdot}$ radicals. The anisotropic nature of silicon etching in a $CBrF_3$ plasma is because of the lower concentration of F_2 and F· present in the plasma and the ion bombardment of Br^+, Br_2^+ and CF_3^+ enhancing the vertical etching process. The high selectivities obtained in etching Si over SiO_2 is probably due to Br_2^+ and Br^+ bombardment, with some CF_3^+ bombardment. $SiBr_4$ is the volatile product formed when Br^+ impinges on the surface weakening Si bonds. Br^+ is not likely to etch SiO_2 but CF_3^+ is believed to.

Paraszcak and Hatsakis (84) have compared the effects of etching silicon when oxygen is added to a CF_4 plasma and CF_2Cl_2 plasma. Using emission spectroscopy they observed overall yields of F· and F_2 concentrations in the CF_2Cl_2 plasma to be much lower than observed in the CF_4 plasma. The etching rate for CF_2Cl_2 was slower than CF_4 for the same amount of O_2 present. In every case the CF_2Cl_2 produced a more aniotropic etching profile. With no oxygen present the CF_2Cl_2 gave a low etch rate but a highly anisotropic profile. The etching of silicon in CF_2Cl_2 appears to be mainly by chlorine ions bombarding the surface, since the bond energy of C-F > C-Cl, chlorine is produced, with C_2F_4 and COF_2 (O_2 addition) being stable products.

Donnelly and Flamm (85) have shown similar effects using CF_3Cl and C_2F_3Cl at low pressures claiming that the rough surface finish after etching is due to ions bombarding the surface.

3.7 Silicon Etching using Carbon Tetrachloride and Chlorine Plasmas

Schwartz and Schaible (86) have used CCl_4 and Cl_2 in RIE of silicon at pressures between 2 mtorr-8 mtorr at an rf frequency of 13.6 MHz. They added argon and helium to the CCl_4 and Cl_2 plasmas which acted as a dilutant in the etching of silicon. They observed that the CCl_4 plasma etched the SiO_2 mask faster than the Cl_2 plasma so that the

Cl_2 plasma gave the best selectivity between silicon and silicon dioxide. The etching of Si in the RIE with CCl_4 and Cl_2 plasmas does not appear to exhibit loading effects. When a negative dc bias was applied to the Si sample, the etch rate increased dramatically in both gases. Both doped and undoped silicon etched in their systems, using a Cl_2 plasma, completely anisotropically, but with highly doped silicon ($\sim 10^{20}\ cm^3$) some lateral etching occurred as shown in fig. 3.16. In the CCl_4 plasma even highly doped silicon etched completely anisotropically and this could have been due to a carbon deposit being sputtered onto the side walls, thus preventing lateral etching. Carbon released from CCl_4 can reduce atomic Cl˙ concentration by recombination and reduce lateral etching.

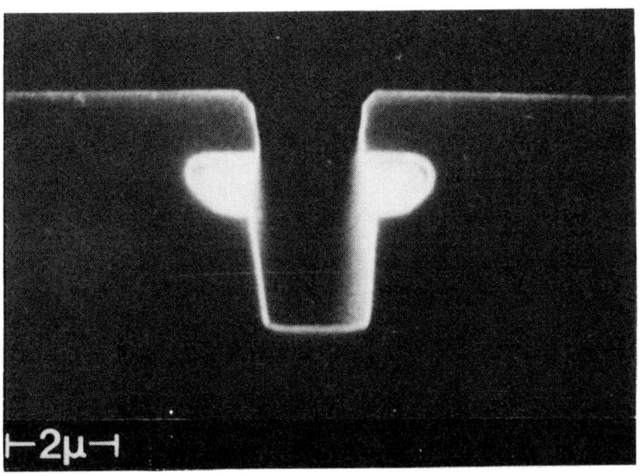

Fig 3.16

Scanning electron micrograph: lateral etching of highly doped N^+ layer from a Cl_2 plasma (ref 86)

Schwartz and Schiable negatively biased the silicon sample and observed an increase in etch rate. This was due to the increase in ion bombardment energy enhancing the etch rate. Since their results show very little observed loading effects coupled with highly anisotropic etching of silicon, this indicates that in their system ion bombardment is the rate determining process for etching silicon, when using Cl_2 and CCl_4 plasmas.

Bruce and Yao (87) however, observed isotropic etching of doped polysilicon at 0.3 torr using a Cl_2 plasma with an excitation frequency of 13.6 MHz. When this was reduced to 100 kHz the polysilicon etching was an order of magnitude faster and completely aniostropic. If the pressure at 100 kHz was increased to 2 torr a small lateral etching component was observed. At 100 kHz the ion bombardment energy is higher than at 13.6 MHz (87) and this should increase the etch rate and etch anisotropically. At the 100 kHz and increasing the pressures to 2 torr one expects an increase in the free chlorine concentration and a reduction in ion bombardment energy, and thus some lateral etching may occur. Bruce and Yao mention the resist erosion problems which can be caused by chlorine bombardment. The selectivity in Cl_2 plasmas between silicon and silicon dioxide are good and hence SiO_2 is very useful for a masking material. When photoresists are used the selectivity between Si and SiO_2 may be reduced since resists are carbonaceous materials and, in a Cl_2 plasma, recombination to CCl_x species can lead to increased etching of the SiO_2. CCl_4 plasmas have a poor selectivity for Si to SiO_2 with etch rates of the SiO_2 being more than five times faster than in the Cl_2 plasma. Most processes for etching polysilicon over SiO_2 have moved from CF_4/O_2 plasmas to CCl_4 and Cl_2 plasmas due to the high selectivity and anisotropic requirements.

Using Cl_2 plasmas the selectivity between polysilicon and SiO_2 can be as high as 70:1. However, if there is any residual SiO_2 film on the polysilicon surface, then etching will not proceed, since SiO_2 is difficult to etch with Cl_2. The wafer first must be dipped into buffered HF solutions to remove any SiO_2 on the surface prior to Cl_2 plasma etching.

Chlorine gas is very corrosive and will increase the maintenance requirements for the plasma reactor's chamber and pumps. Chlorine plasma etching appears to be more successful using lower frequency excitation due to the increase in ion bombardment energy (87). However, Schwartz and Schiable (86) used 13.6 MHz and etched doped polysilicon anisotropically; Bruce and Yao, at the same frequency but at higher pressures, etched doped polysilicon isotropically. This may be due to a difference in doping levels in the polysilicon and the higher atomic chlorine concentrations at higher pressures which will increase the lateral etching component in the Bruce and Yao system.

Chlorine molecules are known to dissociatively chemisorb on single crystal silicon at ambient temperatures and above (88), resulting in a monolayer of Cl atoms bonded to specific sites. Steric factors inhibit the interaction of additional chlorine with the 3-fold surface sites thus preventing the complete co-ordination at the surface silicon atom. The chlorine atom is relatively large and its steric hinderance inhibits $Cl^{\cdot}$ or Cl_2 from gasifying the silicon by the formation of the volatile silicon tetrachloride ($SiCl_4$). Formation of $SiCl_4$ proceeds only at elevated temperatures (89) although some workers' results suggest formation at a lower temperature (90). Observations from ref. 88 have estimated the bond lengths of the Cl chemisorbed at the silicon site to be 1.85A. This is a strong covalent bond, but not as strong as the Si-Cl bond in $SiCl_4$.

In a Cl_2 plasma the polysilicon surface is exposed to Cl^+ and Cl_2^+ ion bombardment, which would enable the Cl to react with the 3-fold axis and help form $SiCl_4$ by disrupting the Si-Si bonds. $SiCl_2$ formed on the surface could also be sputtered away more easily than silicon, rapidly forming $SiCl_4$. Schwartz and Schiable have shown that highly doped silicon can etch with some isotropy in a chlorine plasma, shown in fig. 3.16 Bruce and Yao observed isotropic etching of doped polysilicon only at high excitation frequencies. There are three possible reasons for these observations: (1) There is a larger free radical generation and a lower ion bombardment energy obtained using high frequency rf excitation. Once there is an appreciable $Cl^{\cdot}$ concentration, lateral etching may take place if the silicon is doped since (2) the dopant may produce defects at the surface which allow the free radical $Cl^{\cdot}$ to react with the 3-fold axis producing the volatile $SiCl_4$. Finally (3) dopants may raise the Fermi level which would lower the energy barrier thus making the Si to $Cl^{\cdot}$ bond formation more probable. These will all contribute to isotropic etching of polysilicon. Etching at low rf excitation frequencies with low $Cl^{\cdot}$ concentration and large ion bombardment energies will give high anisotropy for both doped and undoped polysilicon (67). Increasing the pressure of a Cl_2 plasma will raise the adsorption rate and formation of $SiCl_2$ at the surface, thus increasing the etch rate. This will only occur if the ion bombardment energy is above the threshold required for the $SiCl_4$ formation.

e.g. $$SiCl_2 + Cl_2^{\cdot} \rightarrow SiCl_4$$

The formation rate of $SiCl_2$ by the absorption process is fast, as Cl_2 has a high and spontaneous sticking factor on silicon. Increasing the temperature may inhibit the adsorption processes as will the presence of contaminating gas such as oxygen in a system which will compete for silicon sites and terminate etching in a Cl_2 plasma, as was demonstrated by Kravite et al. (91). Mogab (92) obtained similar results on trying to etch Si[100] using Cl_2 plasma and observed a blue film over the wafer. Auger and XPS analyses of the blue film found a high portion of oxygen in what appears to be a redeposited amorphous layer of silicon, containing some $SiCl_2$.

The ion bombardment threshold for Si[100] and polysilicon doped or undoped may all be different. However, etching polysilicon in our Electrotech 317 planar reactor using CCl_4 or Cl_2 at 300 mtorr pressure, is highly anisotropic. There is also no change in anisotropy or etch rate with non-activated, doped or undoped polysilicon. This may be because the ion bombardment energy is well above any threshold.

CHAPTER FOUR

ALUMINIUM ETCHING IN CHLORINATED PLASMAS

4.1 Process Requirements

Films of aluminium (Al) and its alloys are widely used as metal interconnectors in integrated circuits and they are usually etched to form network patterns. As the circuit density becomes greater, the metal interconnector tracks must become narrower and more closely packed. Thus the tracks must be thicker to have sufficient current carrying capacity as their line width decreases. For modern circuit preparation the etching of these thick aluminium films of small track width cannot be achieved by wet etching because of its isotropic etching nature which will produce too much undercutting and line width loss. Hence the interest in using planar reactors for the plasma etching of aluminium based interconnectors producing less undercutting and good line width control.

When aluminium is in contact with silicon, diffusion can occur during high temperature treatment thereby silicon is consumed into the aluminium film which will deteriorate the device. By loading the aluminium film during deposition with 2% silicon, the silicon under the aluminium film will not out diffuse into the aluminium since the Si solubility is satisfied. Also copper (4%) is added to the Al/Si alloy to increase its current carrying capacity when used as an interconnector. However this can pose a problem for etching because the copper can remain on the wafer after etching because it does not form a volatile chloride.

4.2 Processes used for Etching

Fluorinated plasmas cannot be used for etching aluminium or its alloys because aluminium fluoride (AlF_3) is an ionic solid and not volatile under dry etching conditions (vapour pressure 2 torr at 1200 °C). However, aluminium chloride and aluminium bromide (Al_2Cl_6 and $AlBr_3$) are volatile (1 torr vapour pressure at 98 °C and 80 °C respectively) and thus gases containing chlorine and bromine can be used to plasma etch aluminium films. Aluminium will not etch in a barrel reactor since there is a native oxide on the surface, hence ion bombardment is required to remove the native oxide by chemical and physical sputtering.

A number of investigators have etched aluminium films using boron trichloride BCl_3 (93) and carbon tetrachloride plasmas (18) in a planar reactor. Carbon tetrabromide (CBr_4) has not been successfully used because the compound is a solid and does not generate sufficiently high vapour pressure to introduce into a reactor chamber (67). Many workers claim more reproducible etching is achieved when etching aluminium by using a load lock (9) and a cold trap (16) because of the hygroscopic and corrosive nature of the Al_2Cl_6 reaction product formed when using CCl_4 and BCl_3. The load lock prevents atmospheric contamination of the chamber and the cold trap prevents Al_2Cl_6 contaminating the rotary pump oil.

Poulsen et al. (93) first observed the problems which can occur when plasma etching aluminium in a planar reactor operated at 13.5 MHz excitation frequency. Etching with Cl_2 and HCl plasmas resulted in very long induction periods, until the surface oxide (Al_2O_3) was sputtered away from the aluminium surface. The etch rate was then rapid and etching continued even without rf power to the system. The induction period was shortened by using an Ar plasma to physically sputter the oxide before admitting the Cl_2 gas. However, the longer process time deteriorates the photoresist mask and excessive etch undercut which can result when using Cl_2 plasma makes the process unacceptable. Poulsen et al. developed an acceptable process using CCl_4 and BCl_3 plasmas, but claimed an advantage in using BCl_3 as it is a gas at room temperature (93). Fig. 4.1 shows the etch rate of Al as a function of power for different electrode spacings in a BCl_3 plasma.

The increase in etch rate with the increase in electrode spacing could possibly indicate some ion bombardment contribution to the aluminium etching process. An increase in electrode spacing will increase the applied voltage, although Poulsen et al. made no voltage measurements at the electrodes to verify this.

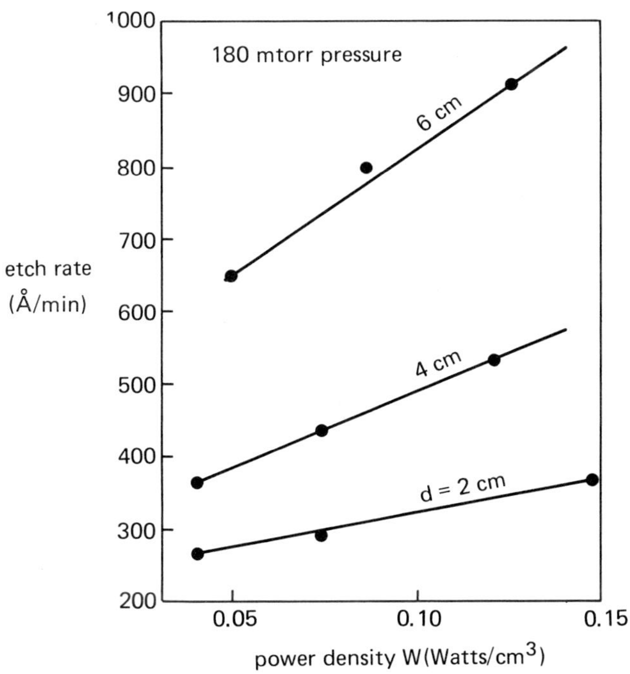

Fig. 4.1.
Etch rate A/min vs rf power for different electrode spacing d in BCl_3 (93).

Schaible et al.(34) also found that an Ar plasma could be used to remove Al_2O_3 prior to the etching of aluminium with Cl_2, Br_2, HBr and HCl plasmas in a RIE. However, by using a CCl_4 plasma, the Ar plasma run prior to the CCl_4 plasma was unnecessary probably because CCl_3^+ ion bombardment was as effective as that of Ar^+ in removing Al_2O_3 from the aluminium surface.

Schaible et al. have achieved aniostropic etching of aluminium using a SiO_2 mask in a plasma containing 2 mtorr CCl_4 plus 8 mtorr Ar. Fig. 4.2 shows the aluminium etch rate versus CCl_4 flow rate at various powers. They also observed that the etch rate of aluminium rose to a maximum value at 125 ^oC and thereafter decreased to a minimum value at 250 ^oC. This may be due to a free radical interaction requiring a surface absorption process which starts to desorb or evaporate at temperatures above 125 ^oC. They report that etching aluminium films appear to suffer from loading effects and is an indication that there is a free radical mechanism which is in limited supply.

Schaible and Schwartz (94) have etched aluminium silicon and copper alloys using a CCl_4 plasma with Ar additions between pressures of 1.3-3 pa. They observed that the lateral etching which occurred was independent of the SiO_2 or MgO mask used. The lateral etching increased with rf power input but films etched on a cooled electrode showed no evidence of lateral etching. The increase in lateral etching with increasing rf power could have arisen from an increase in sample temperature on a non cooled electrode. Thus water cooled electrodes are necessary to obtain minimum lateral etching.

Schaible and Schwartz also observed that aluminium alloy films deposited slowly to form large grain films showed smaller lateral etching than that of small grain deposits, this may be due to lateral etching which occurs along grain boundaries. They also state that there is

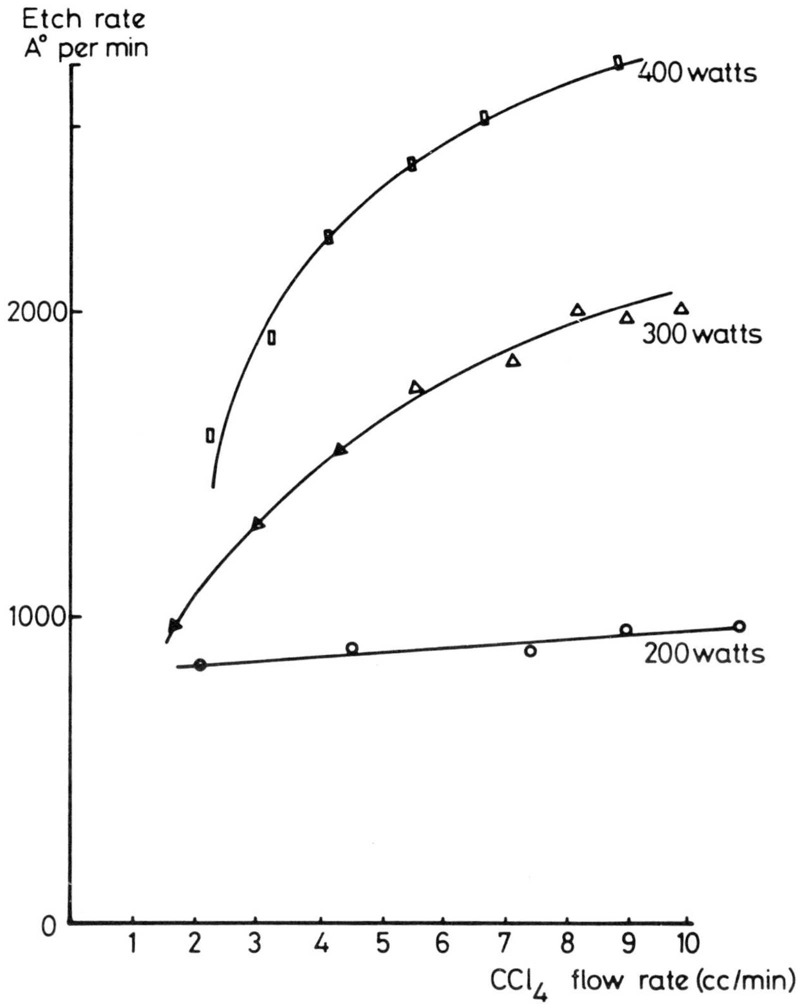

FIG 4.2

ETCH RATE OF Al VS CCl_4 FLOW RATE. ETCH GAS WAS 2MTORR CCl_4 + 8MTORR Ar. ELECTRODE TEMPERATURE, 225°C FROM REF 34

some evidence of film polymerization along the vertical side walls which may result from ion bombardment sputtering the polymer film and depositing on the side walls forming a protective layer against the lateral etching species. This protective layer may be broken down on heating, causing rapid lateral etching which was observed at high sample temperatures. There is no experimental evidence so far reported that polymer formation occurs during etching in a CCl_4 plasma

Curtis and Brunner (95),using a crystal monitor coated with aluminium, observed an initial increase in weight of the aluminium film when etching in a CCl_4 plasma at 0.5 torr, 13.5 MHz and 100 watts of rf power and assumed this to be a polymer film built up on the Al surface and affected the reproducibility of etching. The polymer formation was found to be related to the reactor cleanliness forming thicker films if the aluminium film is allowed to become contaminated before etching.

Hess et al. observed polymer formation on stainless steel (12) electrodes and not on anodized aluminium electrodes during the etching of Al in CCl_4 plasma and trapped the waste plasma products in a cold trap which were then analysed. This analysis revealed an increase in the amount of C_2Cl_4 and other dimerization products when stainless steel electrodes were used. This is probably due to traces of nickel and iron sputtered from the stainless steel electrode which act as a 'Fischer-Tropsch' type catalyst increasing the free radical recombination thus increasing the molecular weight and aiding polymerization.

An investigation by Tokunaga et al. (96) of Al etching properties using CCl_4 and BCl_3 plasmas revealed that the etch rates of aluminium and Al_2O_3 depend on sample temperature as shown in fig. 4.3. Induction times during aluminium etching were always variable in the CCl_4 plasma but with a BCl_3 plasma, shorter and more reproducible etch induction times were obtained. BCl_3 was found to be a better scavenger for O_2 and H_2O contamination because it is a stronger reducing agent than $CCl_2^{:}$ or $CCl_3^{\cdot}$. When small amounts of O_2 were added to BCl_3 plasma the aluminium etch rate increased as more free chlorine was generated and BOCl was formed. With O_2 addition to a CCl_4 plasma the etch rate decreased slightly. At high oxygen levels in the BCl_3 plasma, etching was impaired because B_2O_3 was deposited in the chamber. The BCl_3 scavenging capability means that any contamination will react with the BCl_3 rather than react and passivate the aluminium surface such as O_2 or H_2O leaks into the chamber forming Al_2O_3 on fresh Al surfaces.

The etching characteristics are affected more drastically with the CCl_4 plasma as the aluminium surface can compete as a scavenger in removing the surface contamination. Chamber contamination can also cause variable etching characteristics for aluminium in a CCl_4 plasma.

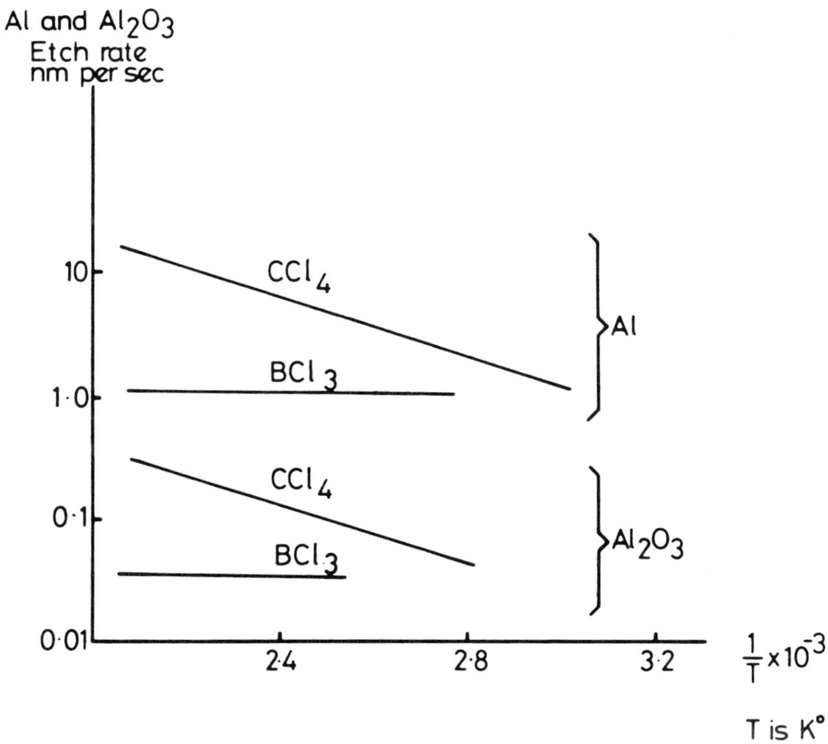

FIG 4.3

ETCH RATE OF Al AND Al_2O_3 IN BCl_3 AND CCl_4 VS TEMPERATURE °K^{-1} (96)

Winkler (9) has shown that atmospheric contamination can be reduced by using a vacuum or nitrogen load lock to the plasma chamber. The introduction of a load lock pre-stage to the system improved etching reproducibility and uniformity. The curves shown in fig. 4.4 were obtained by monitoring the chamber pressure with time and indicated the improvement in the pressure stability when a nitrogen purged load lock was used.

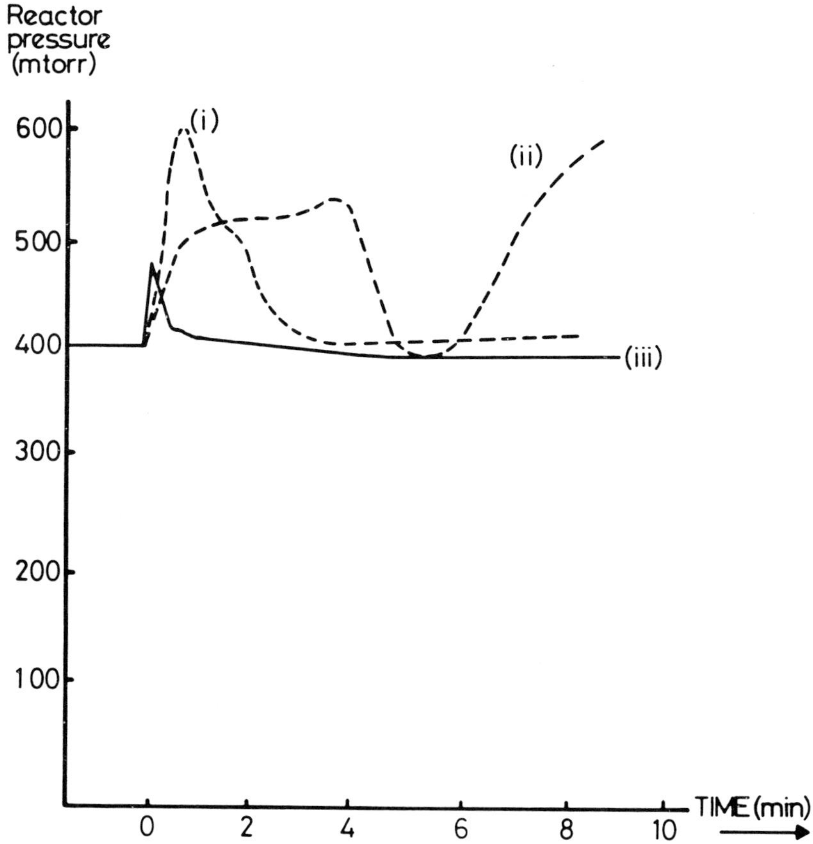

FIG 4.4

REACTOR CHAMBER PRESSURE VS TIME AFTER; (i) CLEANING REACTOR, (ii) TWO WEEKS LATER, (iii) INSTALLATION OF LOAD LOCK, FROM REF 9.

4.3 Aluminium Etching with Silicon Tetrachloride

Sato and Makamura (97) have used a $SiCl_4$ plasma for etching aluminium in a RIE system and observed that silicon dust is left on the sample which can be removed more easily than carbonaceous deposits which can form in CCl_4 plasmas. The $SiCl_4$ plasma used by Sato and Makamura had a slower etch rate than a CCl_4 plasma but produced etched films with less lateral etching compared with CCl_4. The reduction in lateral etching may be from silicon depositing onto the side walls of the aluminium tracks.

The reaction may be where the native oxide is first sputtered away by ion impact and then the aluminium is etcjed by deposition of Si:

$$3SiCl_4 + 4Al \rightarrow 4AlCl_3\uparrow + 3Si\downarrow$$

4.4 Possible Mechanisms in Aluminium Etching

Booth and Heslop (18) have observed the so-called 'bulls eye' effect when etching aluminium films in a CCl_4 plasma. In this case the aluminium films on Si wafers are etched from the edge to the centre. They have also shown the effects of wafer loading in a planar reactor where the etch rate is reduced as the aluminium wafer area is reduced. Both these effects indicate that generation of free radicals for reaction is limited.

Walters (98) has shown that there is an increase in lateral etching of aluminium films when using an SiO_2 mask instead of a photoresist mask in a CCl_4/Cl_2 plasma. The lateral etching is reduced to zero when SiO_2 masks are in close proximity with photoresist masks, this effect is illustrated in fig. 4.5. It is believed that the photoresist present may either consume the free radicals that are responsible for undercutting or it is sputtered away and redeposited onto the side of the aluminium tracks thus reducing lateral etching.

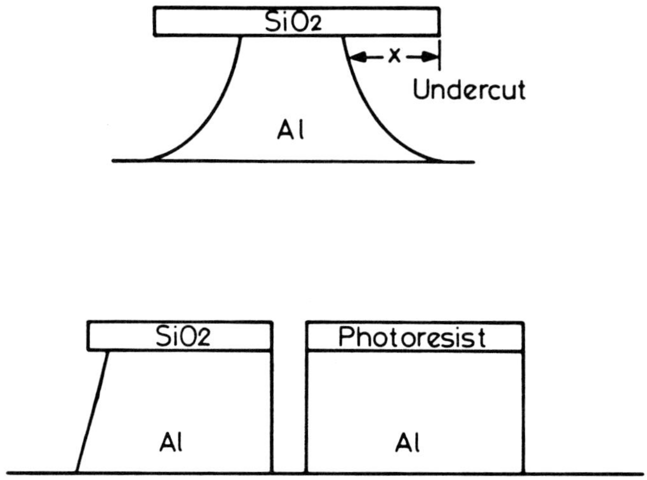

FIG 4.5

UNDERCUT IN ALUMINIUM FILMS WITH DIFFERENT MASK MATERIALS.

The fact that aluminium etches with some degree of anisotropy indicates that ions also play a role in etching aluminium and reactive ion beam etching (39) using Cl^+ ions has been used successfully for etching aluminium and its alloys, because it has an anisotropic etching component without any detected undercutting. Although etching Al and its alloys by RIBE normally has a slower etch rate than in plasma etching, an advantage is that the etched films do not appear to suffer from post etch corrosion. In plasma etching and RIE of Al films the free radicals used have strong chemisorption properties and if left on the aluminium tracks will absorb water and corrode the metal. These chemisorbed radicals are not removed in a vacuum and when the etched Al wafer is removed from the chamber it must be washed immediately to prevent post etch corrosion.

Plasma etching of aluminium appears to be dependent on free radical interactions as well as a possible ion impact contribution, but the fact that aluminium is so reactive chemically makes it susceptible to lateral etching. However, some etching processes appear to be completely aniostropic, which may arise from:

(i) Formation of Al_2Cl_6 which does not readily leave a cool substrate. This compound may be sputtered or helped to volatize by ion bombardment which, if directional, gives anisotropic etching. If the side of the track elements were covered with Al_2Cl_6 this would then prevent further lateral etching where ion impact does not occur, such an effect would occur only if the substrate temperature was low.

(ii) Ion bombardment may also sputter photoresist or polymer onto the side walls of track elements, giving a coating of carbonaceous material preventing lateral etching by free radicals.

(iii) In low pressure plasma systems in particular with BCl_3, the free radical concentration is low and thus reduces any lateral etching. Most manufacturers are now operating in the 10-20 mtorr range with RIE systems using BCl_3 plasmas and claim no undercutting in etched Al films.

CHAPTER FIVE

THE PLASMA ETCHING OF III/V COMPOUND SEMICONDUCTORS

5.1 Why III/V Compound Semiconductors?

Group III/V compound semiconductors are the most recent materials used for high frequency device operation as they have a higher carrier mobility than silicon (99). They have also been developed for laser devices and light detection (100).

5.2 Process Considerations

The compound formation of a III/V system such as gallium arsenide (GaAs) makes it particularly difficult to plasma etch because the surface stoichiometry can be affected by the variation in the volatility of the reaction products of Ga and As. With plasmas containing fluorine, such as CF_4, the surface of GaAs becomes gallium rich as there is an initial loss of AsF_3 which is far more volatile than GaF_3. The GaAs surface can also be subjected to ion bombardment which can cause heating and melt the gallium rich surface forming droplets (66). Most workers tend to use plasmas containing chlorine because both $GaCl_3$ and $AsCl_3$ are volatile reaction products.

5.3 Plasma Conditions used for Etching III/V Compounds

Smolinsky et al. (101) have etched GaAs and its oxide in a number of chlorine-containing plasmas. They reported that GaAs etched most rapidly in the following: chlorine Cl_2, phosgene $COCl_2$, carbon tetrachloride CCl_4, carbon difluodichloride CF_2Cl_2, phospherous trichloride PCl_3 and hydrogen chloride HCl, with etch rates of up to 5 μm per min at 0.5 torr and in a supply frequency of 13.5 MHz using a planar reactor.

The oxide of GaAs etched in all of these gases apart from Cl_2 and $COCl_2$. However, the oxide (Ga_2O_3/As_2O_3) etch rate was much slower than that of GaAs. Other gases used by Smolinsky et al.such as perchloroethene C_2Cl_4, carbon dichlorodibromide CCl_2Br_2, chloroform $CHCl_3$ and phosphorus hydride PH_3 appears to deposit films apparently before etching could occur.

All of the plasma etchants studied roughened the GaAs surface during etching. The most severe was Cl_2, which had the fastest etch rate, and hydrogen chloride gave the lowest amount of roughening and the slowest etch rate. An HCl plasma would generate the smallest concentration of free $Cl^{\cdot}$ because of the high recominbation rate of free chlorine with hydrogen. The $Cl^{\cdot}$ species appears to be an active component in the etching of GaAs and the $Cl^{\cdot}$ concentration would decrease with the rise in hydrogen concentration because the heat of formation of HCl is high (89).

Asaro et al.(102) observed that a Cl_2 and boron trichloride BCl_3 mixture at 0.15 torr and at an operation frequency of 27 MHz etched GaAs anisotropically but with higher concentration of Cl_2 some undercutting was observed. Birton et al.(103) obtained similar results at a high pressure of 0.3 torr and a lower supply frequency of 55 kHz when using CCl_4/Cl_2 mixtures.

Caldren and Rentschler (104) achieved anisotropic etching of indium phosphide (InP) with a Cl_2/O_2 mixture in a reactive ion etching system at 5 mtorr pressure. At this low pressure and at high power inputs, etching tended to give rougher surfaces.

Chang and Davack (105) used a hydrogen plasma for etching Ga_2O_3/As_2O_3 to expose the GaAs undersurface giving a selectivity oxide to GaAs of 2:1. When an H_2 plasma is used to etch Si and SiO_2 by forming SiH_4, the etch rate is slow (67). However, with a H_2 plasma the oxide of

GaAs etch rate is fairly fast reaching 600A per min at 400 watts, 0.3 torr for a supply frequency of 30 MHz. Chang and Darack have pointed out that the electron temperature T_e of their system must be greater than 10 eV at high frequency (30 MHz) as the H_2 molecule has a very high ionization potential (19) and small cross-sectional area for collision.

Flamm et al. (106) have etched GaAs and silicon in a plasma containing bromine using his radial flow parallel plate reactor. The etch rate at 100 watts, 2 MHz applied rf frequency, 0.3 torr pressure was 1750 Å/min for GaAs. When InP was etched under the same conditions, the etch rate was very low (50 Å/min). This is probably because GaAs has the lower activation energy for surface reaction in this process. The surface topography after etching InP and GaAs appears to have a strong dependence on the surface temperature becoming smoother with increasing temperature.

Flamm et al. (107), using a Cl_2 plasma with optical diagnostics, observed that GaAs etching exhibited little dependence on applied frequency at constant input power and pressure (0.3 torr). However, the InP etch rate increased with decrease in the supply operating frequency indicating that the etching of InP is a more ion dependent reaction (higher activation energy). The applied rf potential increases for the same power density as the frequency decreases and thus the ion energy increases with decreasing operating frequency. Flamm et al. do not indicate if the etching aniostropy changes with frequency for InP. As the etching of InP in a Cl_2 plasma appears to be the most ion dependent the etching characteristics should be more anisotropic compared with GaAs.

CHAPTER SIX

OPERATING FREQUENCY AND THE PLASMA

Schwartz and Schaible (86) observed, during the etching of aluminium in CCl_4/Ar plasmas, a slight increase in etch rate when the frequency was decreased from 27.12 MHz to 13.56 MHz. Coburn and Kay (55) observed similar changes in etch rate when decreasing the operating frequency. The reason proposed was that due to the increase in network efficiency at lower frequencies, the power dissipation into the plasma improved. Later, other workers (113,116) observed changes in ion impact energies and free radical concentration as the applied operating frequency was changed, which may affect the etching properties of the plasma.

Plasmas are very complex systems, any single parameter change (such as pressure) may affect many other parameters. Some workers have attempted to analyse plasmas by using electrical circuit analogy.

6.1 The Equivalent Circuit

Koenig and Maissel (108) have provided a simple equivalent circuit for rf discharges, and although proposals by others differ, the basic concepts are similar and are shown in fig. 6.1.

The dark space sheaths are primarily capacitive due to the positive ion charge in the sheath. This arises from the negative surface charge on the electrode or grounded adjacent surface. The presence of the blocking capacitor results in charge storage, and although the ion and electron currents differ, their total charge transfer of each half cycle is equal and balanced. The ion sheath characteristics can change considerably with gas pressure due to recombination and collisions. These are a function of number density and collision cross section, i.e. mean free path.

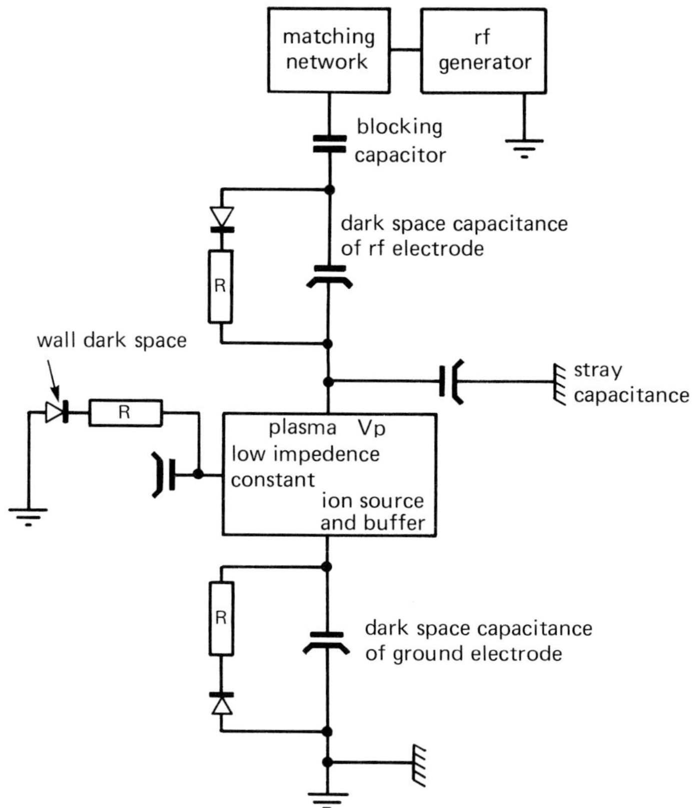

Fig. 6.1 A simple diagram for the equivalent circuit of an rf discharge

6.1.1 Matching network

Matching networks are used in high frequency discharges (above 1 MHz) to match the impedance of the load (in the case of a vessel containing electrodes) to the rf generator output impedance. When the impedances of supply and load are matched, i.e. are equal, the power transferred by the generator to the discharge is at a peak value.

By using a dc counterpart theorem, the power (P) from the r.f. generator will vary with the value of R for the plasma load system

$$P = \frac{E^2R}{(r+R)^2} \quad .$$

The maximum value of P is obtained by differentiation

$$\frac{dP}{dR} = \frac{E^2(r+R)^2 - 2(r+R)\ E^2R}{(r+R)^2} \quad .$$

This differential has a zero value for R = r which is for maximum P. Thus to dissipate maximum power into the plasma load (R) we must match the resistance of the plasma load to the resistance of the rf power supply (r). A simplified diagram of a typical matching network is shown in fig. 6.2.

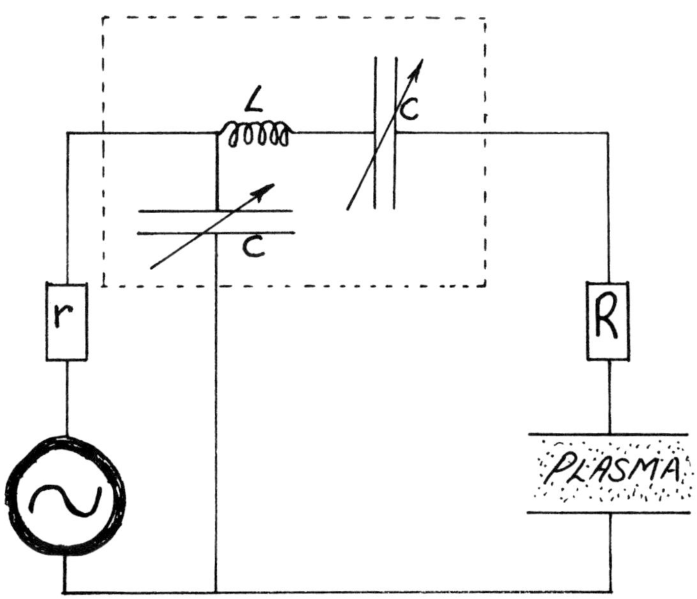

Fig 6.2

Simplified diagram of a typical matching network

Using a matching network with known component values, Logan et al.(110) were able to calculate values for the conductance and capacitance of an rf discharge. They pointed out, however, that the losses in the matching network must be small. The values of conductance and capacitance for the discharge were deduced from measurements on the matching network after termination of the discharge. They also further observed an increase in the conductance and capacitance of the discharge with an increase in the pressure of the discharge. This would arise from the space charge depth decreasing, and the plasma region having a higher balanced charge density with rise in pressure.

6.2 Applied Potential and Conductance Changes in Plasmas and the Corresponding Changes in Etching Properties

Norström (111) observed that the conductance of a discharge was dependent on the applied frequency and pressure as shown in fig. 6.3.

Norström could not sustain the discharge below the pressure points 'P' at a particular frequency, showing that high frequencies are needed to sustain discharges at low pressures.

Using Langmuir probes, Norström (112) also observed, for the low power plasma (20 watts, 10^{-1} torr Ar), an increase in the plasma potential to ground as the applied rf frequency was decreased, as shown in fig. 6.4.

Fig. 6.4 shows that, apart from having an unusual negative plasma potential to ground at 14 MHz, the probe current saturation values rose with increasing frequency. This indicated that there may be additional ionization events at higher frequencies giving larger ion densities and thus larger probe currents.

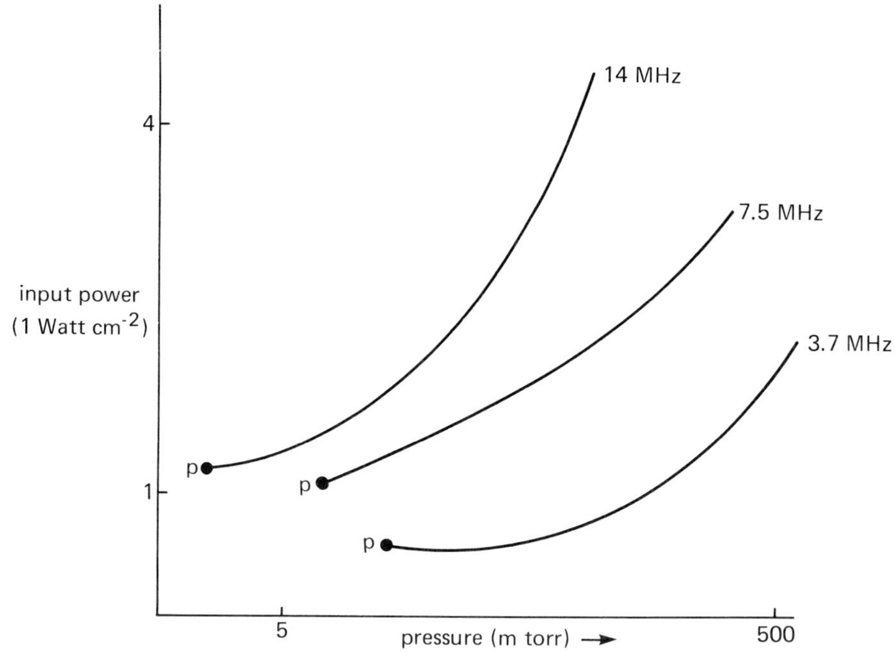

Fig. 6.3

Conductance as a function of pressure (111) using a planar type of reactor at various rf frequencies.

Norström demonstrated that an insulating material covering the metal wire in a Langmuir probe was found to affect the discharge characteristics. This was because the extra ejection of secondary electrons from an insulating material increased slightly the conductance and the electron density of the plasma. Insulating material was deposited on the probe from the sputtering of the target material which affected the discharge characteristic. This was reduced by coating the target with colloidal graphite which has a low sputtering yield (and should not affect the ionization conditions of the plasma by secondary electron emission). The introduction of the probe did not affect the plasma characteristics providing the target was coated with graphite, thus reducing sputtering from the target onto the probe.

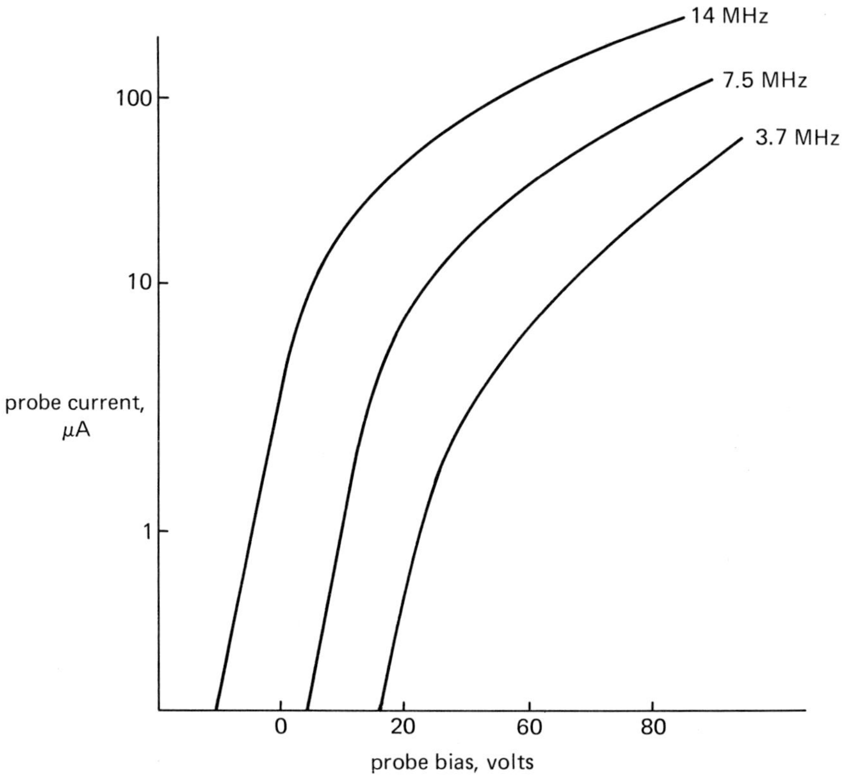

Fig. 6.4.

Probe current as a function of probe voltage at different rf frequencies using a planar type reactor at constant power (112).

Fig. 6.4 also shows that the slope of the I/V characteristics appear to be similar and hence there is no significant change in electron temperature with applied frequency.

Using a mass spectrometer, Bruce and Reinberg (113) studied the etching of doped and undoped polysilicon. They used a chlorine plasma at 0.3 torr pressure at between 1.5 kHz to 27 MHz in a planar reactor connected by a pin hole to a differentially pumped chamber containing the mass spectrometer. They observed aniostropic etching over the frequency range but found that the etch rate of Si increased slightly to a maximum at 500 kHz and then decreased rapidly as the frequency was increased above 500 kHz at constant power and pressure.

This change in the silicon etch rate with frequency, coincided with the change in applied potential to the rf electrode. The increase in silicon etch rate was accompanied by an increase in the applied potential to the rf electrode and is shown in fig. 6.5.

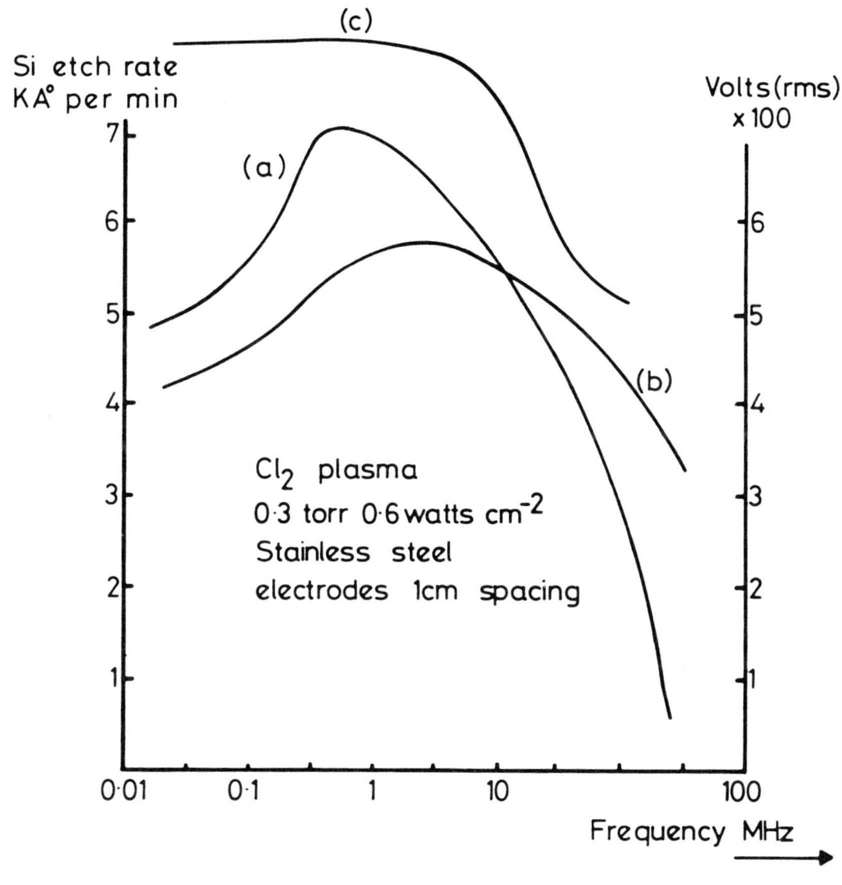

FIG 6.5

(a) VARIATION IN Si ETCH RATE WITH EXCITATION FREQUENCY ARISING FROM (b) VARIATION OF APPLIED VOLTAGE TO RF ELECTRODE WITH EXCITATION FREQUENCY (c) VARIATION IN ION ENERGY AS A FUNCTION OF EXCITATION FREQUENCY. (113)

The rf power was coupled to the reactor with different circuits at high and low frequency. At high frequencies (above 2.0 MHz) a matching network was used while at low frequency below (1.0 MHz) a transformer was used. In both cases Bruce and Reinberg used a choke to provide a dc path to ground for the live electrode. To ensure that the power delivered to the plasma was the same in both frequency regions, the emission from a 75W light bulb was first calibrated using dc and then they used the bulb to calibrate the rf power meters through the appropriate coupling circuits. The mass spectrometer was attached below the ground electrode and differentially pumped so that it could sample the ions from the plasma. An ion energy analyser was also incorporated in the system so that the mass and energy of the ions could be determined.

Measurement showed a rapid ion energy reduction as the operating frequency increased above 500 kHz. The monitored Cl_2^+ concentration in Bruce and Reinberg's experiment was independent of frequency but the Cl^+ concentration decreased as the frequency was raised as shown in fig. 6.6. The rise in ion energy at low frequencies coincides with the rise in the rf potential at the rf electrode. The rise in the rf electrode potential will increase the plasma potential which will increase the ion energy traversing the dark space sheath thus increasing the silicon etch rate.

The decrease in the Cl^+ current shown in fig. 6.6 is probably due to a reduction in the electron temperature (Te) since Cl^+ has a high ionization potential (15 eV) whereas the Cl_2^+ has a lower ionization potential (11 eV) and thus remains constant. The reduction of Cl^+ concentration by possible reduction in Te as the operation frequency is increased, indicates a larger concentration of high energy electrons at low frequencies (high plasma potentials). This change in electron energy distribution could have other effects such as the free radical composition, which may influence etching. However, the Si etching is very ani-

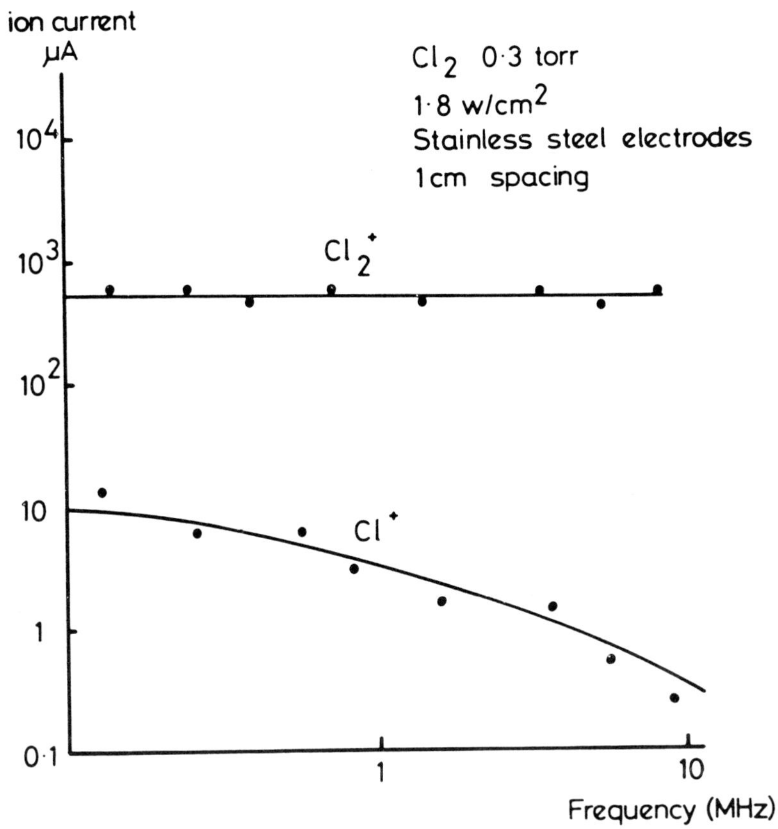

FIG 6.6

ION CURRENTS FOR Cl^+ AND Cl_2^+ AS A FUNCTION OF APPLIED FREQUENCY. (113)

sotropic in Bruce and Reinberg's experiment at both high and low frequencies, thus it appears to be an ion impact induced process.

Ions may not only increase their energy by increase in applied rf electrode potential, but also respond to the rf field at lower frequencies more efficiently, and consequently the energy transfer is larger as shown in fig. 6.5c. At higher frequencies a relatively slow ion moves through the rapidly changing dark space sheath and only achieves a time average energy. The small peak observed in silicon etch rate in fig. 6.5a at 500 kHz is because of an increase in rf electrode potential probably caused by the increase in impedance of the system at that particular frequency.

If a discharge is to be operated at constant power input with reduction in the excitation frequency, then a higher potential is developed at the rf electrode since the impedance across the dark space capacitance increases. Thus the potential across the dark space must increase until the ions can traverse the dark space sheath during part of the cycle.

At high frequencies (13 MHz) the discharge responds, and couples more easily through the dark space sheath capacitance, thus a lower rf potential is needed to traverse the ions through the sheath and maintain ionization. At 13 MHz a phase shift appears between current and voltage which has been reported to be as large as $\pi/2$ (113). At low frequencies, 100 kHz such a phase change does not occur and is resistive to charge flow and this makes it more difficult to couple through the dark space sheath capacitance and hence larger voltages are required. The author (15) has observed small phase shifts of $\sim \pi/8$ at 380 kHz for Ar, CCl_4 and O_2 discharges. Fig. 6.7 shows photographs of voltage and current traces at 100 kHz and 13.56 MHz excitation using a planar type reactor from ref. 113.

The fact that higher rf applied potentials are needed to sustain a plasma at lower frequencies can result in a higher negative dc bias of a capacitively coupled (blocking capacitor) rf electrode (V_s) and also a higher plasma potential to ground potential (V_p). These potentials V_s and V_p will produce a space charge sheath at both electrode surfaces.

At the peak of the rf cycle, the electron field through the space charge sheath will also be at the peak intensity, consequently secondary electrons emitted from the electrode surface can undergo sufficient acceleration to ionize gas atoms or molecules within the adjacent space charge sheath resulting in a multiplication process. The start can be observed in fig. 6.7 showing the hump in the current trace in the 100 kHz excitation. This effect has not only been observed by Bruce and Reinberg but others (114,115,121 and 15) using high pressures and low operating frequencies. Under these conditions the electron mean free path is equal to or smaller than the space charge region. In addition the electric field in the space charge region is large enough to accelerate the secondary electrons (produced by ion bombardment) to ionization

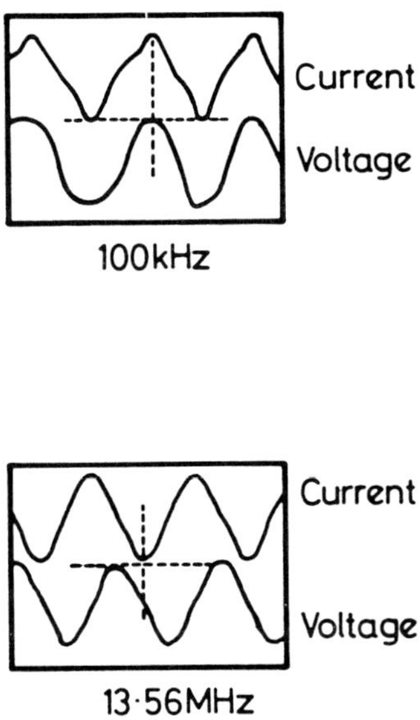

CL_2
0·3 torr
1·8 W/cm^2
stainless steel electrodes
1cm spacing

FIG 6.7

OPERATING VOLTAGE AND CURRENT WAVEFORMS FOR 100kHz AND 13·56MHz EXCITATION FREQUENCY. (113)

energies within the electron mean free path distance (from the electrode surface). Such conditions will cause ionization of neutral molecules within the space charge region by electrons leaving the electrode surface and is most probable when the electric field across the space charge is a maximum, i.e. at the peak of the applied rf voltage. The ions formed in the space charge region will accelerate and bombard the electrode surface producing more secondary electrons, thus a multiplication in current will be observed. This will be discussed further in chapter 8 with experimental evidence.

Systems that generate large dc negative biases on rf electrodes may be approaching (avalanche) multiplication conditions for a large part of the cycle. Holland (121) has concluded that this dc ionization can supplement rf excitation i.e. electron oscillation, with secondary electrons being released by ion impact at the negative biased electrode.

Bruce and Reinberg (113) have not only observed the increase in silicon etching with an increase in ion bombardment energy (showing ion enhanced etching) but also the spread of the ion energies for Cl_2^+ and Cl^+ at high and low frequency, as shown in fig. 6.8. Bruce and Reinberg (113) removed any dc bias formed on the rf electrode by using a choke in their system, and assumed that the sheaths formed on the rf and ground electrodes will be approximately the same. Hence, sampling the ion energy distribution at the ground electrode should be similar to that at the live rf electrode. Fig. 6.8 shows that the ion energy distribution is much larger for the Cl_2^+ and Cl^+ ions at 100 kHz than at 13.7 MHz. This is because ions crossing the sheath at low frequencies can traverse it in a small part of the rf cycle. This will produce a broad energy distribution, because some ions encounter a higher electric field than others. The ions traversing the space charge sheath at high operating frequencies take several cycles to cross the sheath, giving a time-average energy for the ions. This will produce a narrow ion energy distribution compared with low frequency operation. This is an oversimplified concept for high frequency operation, since it assumes that the plasma ionization conditions remain the same and that ion bombardment plays a major role in sustaining plasmas at high frequency. It will be shown later in this chapter that there is some evidence to suggest that there are other mechanisms apart from ion bombardment which sustain plasmas at high frequency.

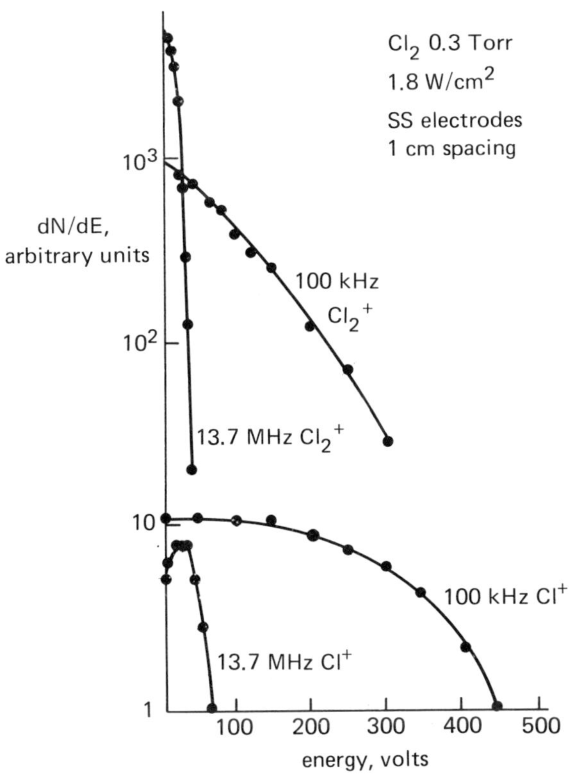

Fig. 6.8.

Ion energy distribution for Cl^+ and Cl_2^+ at 100 kHz and 13.7 MHz excitation frequency from ref 113.

Fig. 6.8 shows that Cl_2^+ has a smaller distribution than Cl^+ both at 13.7 MHz and at 100 kHz, although this is less obvious at 100 kHz. The difference in energy distribution with ion mass was investigated by Coburn and Kay (32). Their work at 13.5 MHz, 75 mtorr using an electrostatic analyser and mass spectrometer concluded that the energy spread (ΔE) of the ion distribution was proportional to $M^{-\frac{1}{2}}$ where M is the mass of the ion considered. Thus large ion masses will have a smaller ion energy distribution. Hence Cl_2^+ will have a smaller energy distribution than Cl^+ probably due to the larger ion mass which will have a slower response to changing electric fields, thus contributing to more time average effects. The energy spread will also be affected by current/voltage and pressure conditions in the discharge particularly with ions colliding with neutral atoms whilst traversing the space charge sheath.

Bruce (116) measured the etch rates of aluminium and silicon as a function of excitation frequency at 0.3 torr pressure using a planar reactor. Al was etched with a plasma consisting of 1:1 ratio of Ar/CCl_4 gases at 75 watts rf power and Si was etched in CCl_4 plasma at 25 watts rf power. He observed an increase in the Al etch rate and a decrease in Si rate as the excitation frequency was increased.

During the plasma etching Bruce analysed the CCl_4 fragmentation changes with frequency variations, using a mass spec system. He also observed the fragmentation changes with variation in rf power. Bruce found a larger Cl_2^+ ion concentration than CCl_x^+ in the plasma when increasing the operating frequency. He points out, however, that the main reaction differences with varying the frequency, for Si and Al etching, is in the vast number of free radicals produced compared with ions. Silicon etches faster at lower frequency since the ion energy bombarding the surface increases (113). Aluminium however may be etched faster at higher frequencies due to the possibility of more polymer and free radical formation occurring at higher frequencies.

At higher frequencies the C_2Cl_2 concentration in the plasma appears to increase with a decrease in mono carbon species type $CCl_x^{\cdot}$, all indicating an onset of polymerization. This condition may help in the surface etching of aluminium.

Bruce points out that the decrease in silicon etch rate at higher frequency may also be due to polymer formation as well as a decrease in ion bombardment energy. The qualitative measurement of a radical species such as CCl_x, or even chlorine products, is difficult since these species are also products of CCl_4 cracking from the mass spec.

Flamm and Donnely (117) have made similar investigations, to Bruce and Reinberg, on Cl_2 plasma etching of doped and undoped silicon with varying r.f. operating frequency. The difference is that Flamm and Donnely also monitored $Cl^{\cdot}$ concentrations using emission spectroscopy as well as monitoring the Cl^+ & Cl_2^+ concentrations (with a mass spec system) and the peak to peak potentials developed at the live electrode. Their results are shown in fig. 6.9. The etch rate of silicon in their system also decreased as the frequency was increased as did the peak to peak voltage on the live electrode, which is in good agreement with Bruce and Reinberg's work (113). The reduction of Cl^+ concentration at higher frequencies could be due to a decrease in electron temperature (T_e). Although high frequencies tend to give a lower rf electrode potential, they can produce large negative dc offsets to ground in some systems (118). Hence the ion bombardment energy conditions may differ at each electrode surface. These effects can change the etch rate even in apparently symmetrical and low frequency plasma systems (15), where a small potential sheath difference can change the etching rate of polysilicon in a CCl_4 plasma.

The work of Bruce and Rynberg (113) and Flamm and Donnely (117) was concerned with the etching of samples on the ground electrode and monitoring the ion energies which have traversed the ground electrode sheath. The reduction in voltage appearing at the live rf electrode at higher frequencies (shown in fig. 6.9), and the probable reduction in

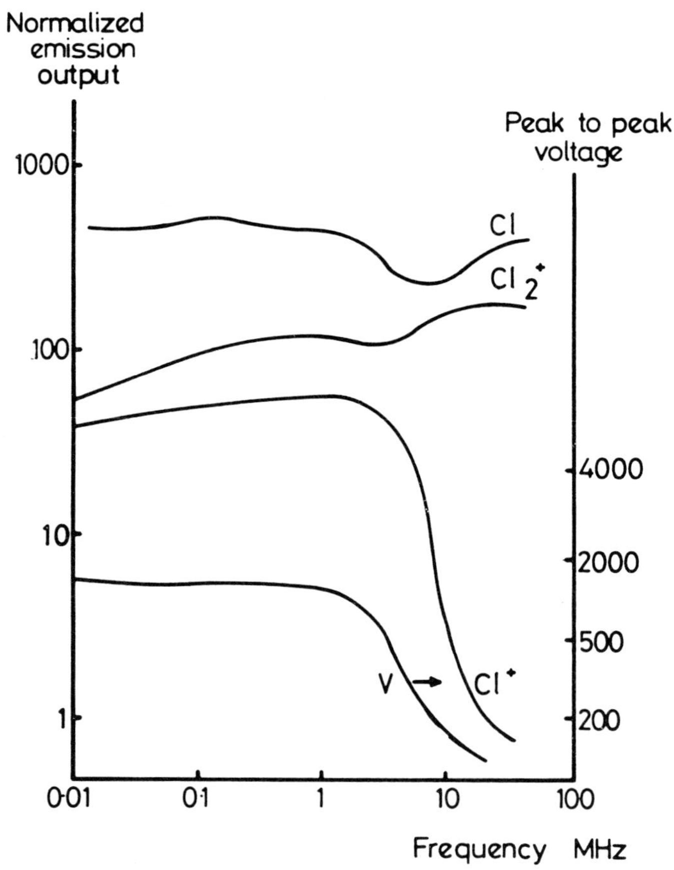

FIG 6.9

NORMALIZED EMISSION FROM $Cl^{\bullet}$, Cl_2^+ AND Cl^+ AS A FUNCTION OF EXCITATION FREQUENCY. PEAK TO PEAK VOLTAGE IS ALSO SHOWN IN THE LOWER CURVE FROM REF 117

plasma potential, will reduce the ion bombardment energy at the ground electrode. The reduced ion bombardment energy produces a reduction in etch rate at the ground electrode.

The change to a higher frequency will also change the type of species produced due to the change in ionization and sustaining mechanisms. A change in plasma potential may change the electron temperature and hence the ionization probabilities. At higher frequencies (13.5 MHz) N-type polysilicon showed a higher etch rate in a Cl_2 plasma, providing that the dopant is activated. There is no change in Si etch rate with P type dopants activated or unactivated (119), as shown in fig. 6.10. Systems operating at lower frequencies and at similar Cl_2 pressures do not show any change in the etch rate of polysilicon with doping, activated or not.

High operating frequency plasmas have a smaller ion enhanced mechanism, (larger free radical concentration) thus free radicals may make more of a contribution to the etching process. This contribution, although still aniostropic, may be electrically activated by the effects of N type dopants on the silicon Fermi level.

The properties of plasmas are not fully understood, particularly for high pressures at low operating frequencies. This may be because workers originally investigated low pressure high frequency rf and dc plasmas for sputtering applications. More recently plasma etching has been investigated at high pressures and generally lower frequencies.

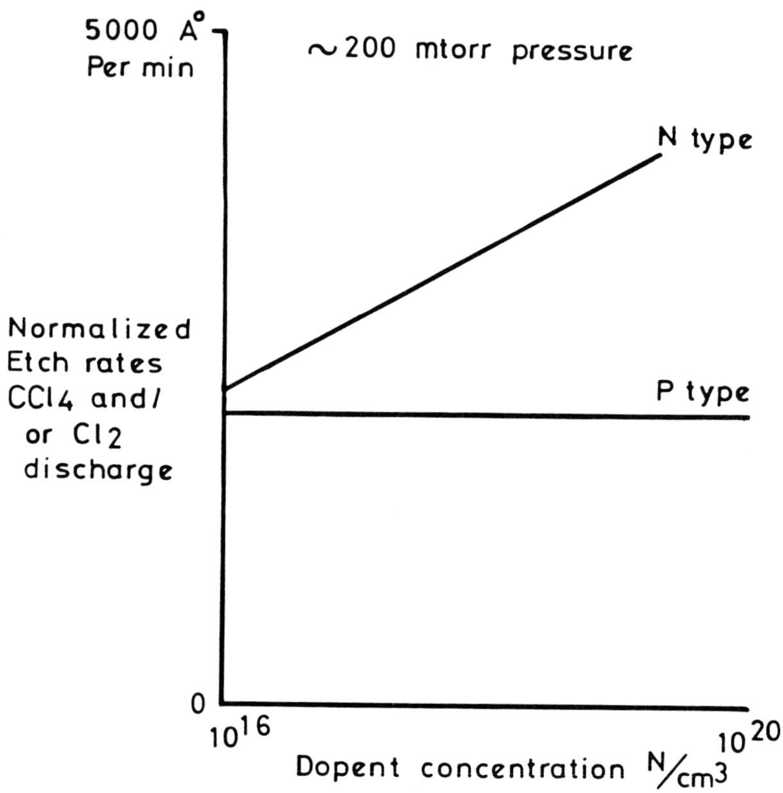

FIG 6.10

SILICON ETCH RATE AS A FUNCTION OF TYPE OF DOPING AND CONCENTRATION AT 13·5 MHz IN A PLANAR REACTOR USING A Cl_2 PLASMA AT 200 MTORR PRESSURE (119)

6.3 Sustaining Plasmas at Low and High Frequencies

Koenig and Maissel (108) observed, in keeping with the general observation, that an increase in operating frequency of the rf supply allows a plasma to be sustained at lower pressures as shown in fig. 6.11.

Norström (111) could not measure the capacitance or conductance of the plasma at low frequency and low pressure due to the difficulty in sustaining the plasma. This explains the lack of information on the capacitance and conductance of plasmas at low pressures and at low operating frequencies.

Any electron energisation in the oscillating field is small at low frequency, where any free electrons will be lost to the walls by drift. This effect combined with a low ion yield at low pressures, since there is an increase in the mean free path, results in a weak discharge, or failure to excite a discharge with low frequencies.

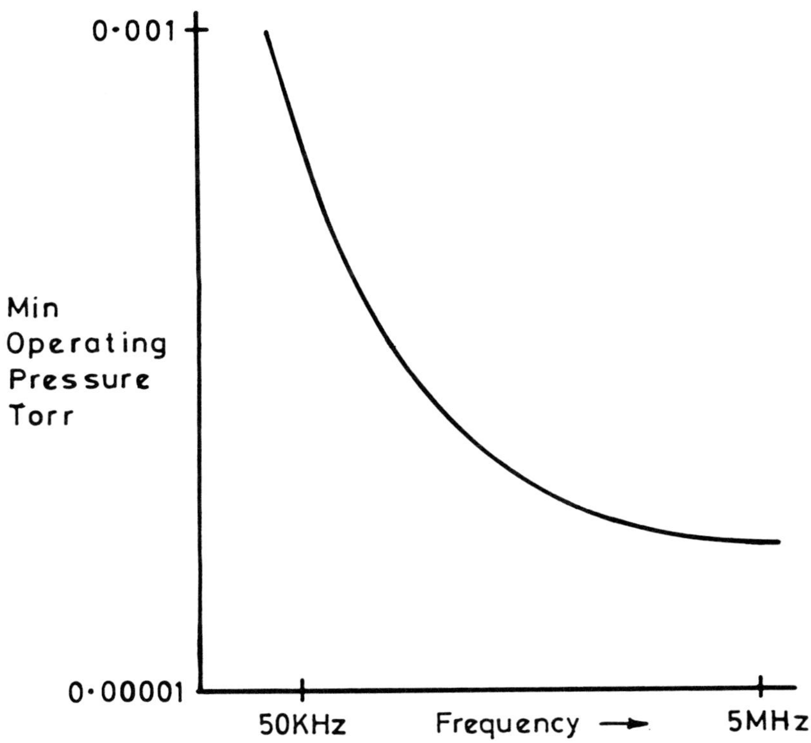

FIG 6.11

MINIMUM OPERATION PRESSURE AS A FUNCTION OF EXCITATION FREQUENCY FROM REF 108

The capability of sustaining a plasma at low pressures and at high frequencies indicates that there is an additional source of ionization other than the secondary electrons ejected from the electrodes by ion bombardment. The contribution from secondary electrons producing ionization may change and to some extent the sheath width due to the change in operating frequency and pressures.

The work of Koenig and Maissel (108) proposed that at high frequency operation secondary electrons are not totally responsible for maintaining the discharge. Thus large voltages are not required for sustaining the discharge. The reduction in rf operating voltages on the live electrode at higher frequencies has been observed by other workers (113). The main source of ionizing electrons, which maintain the discharge, are electrons in the discharge, which are energised by the oscillating rf field. These electrons have elastic collisions with gas atoms in the discharge. These eventually acquire sufficient energy to cause ionization.

The question is how do electrons oscillating at these higher frequencies acquire enough energy to ionize atoms?

In the plasma the presence of a space charge sheath adjacent to the electrode surface will partially screen the plasma from the changing rf fields. Most of the potential will be dropped across the dark space and only a small field will be present across the plasma (28). This is probably true for high pressures (above .05 torr), but at low pressures the impedance of the plasma may be larger and hence so will the electric field across it.

At high frequency and low pressure the electrons may gain energy and change direction with the changing electric field. However, the electric field across the plasma is probably still too low to accelerate electrons to ionizing energies. But the electrons may oscillate in the electric field and gain energy by elastic collisions at an appropriate time with respect to the phase change of the electric field. By these elastic collisions, at appropriate times, electrons can continue to increase in energy to the point that they can ionize atoms thus contributing to the ionization processes in the plasma (120).

There have been several suggestions how the extra ionization at high frequency is caused, and how the electrons accumulate energy by oscillating in the electric field (121,122).

Additional ionization in plasmas can arise from secondary electrons which are accelerated across the target and wall sheaths. The energy of the electrons emitted into the plasma will depend on the field in the space charge sheath changing at the right time. The collision of an energetic electron with the walls or electrodes can produce secondary electrons which enter the plasma to continue ionization processes. A model similar to this has been proposed by Holland et al. (121) and Gill and Von Engel (122). Gill and Von Engel suggest further ionization can occur if an electron, starting at one electrode or wall with a velocity in a suitable phase with the applied field, crosses the vessel (in one or an odd number of $\frac{1}{2}$ cycles) hitting the other electrode or wall (opposed by a weak electric field at the time of impact) fast enough to release on average more than one secondary electron which will then help sustain the plasma. Electron bombardment onto surfaces will generate more secondary electrons than ion bombardment (19) since they are more efficient and thus will readily contribute to sustaining plasmas.

Keller and Pennebaker (123) reject the possibility of further ionization by fast electrons from walls or electrodes and conclude that ionization was due to the large rf currents flowing through the plasma.

The electrons in the Keller and Penebaker model gained energy by 'surf-riding' on the oscillation edge of the space charge sheath. The applied rf voltage causes the space charge sheath adjacent to the electrodes to grow and decay. If an electron moves with a velocity V_e and collides with the space charge sheath boundary, which was in a period of growth and moving at a velocity of V_d, the electron will then recoil. At high frequency the velocity (V_d) can be appreciable, even if the space charge electric field change was small. The electron, on collision with the space charge boundary, regards this as a solid object and thus will recoil in the opposite direction with a new energy proportional to $V_e + 2V_d$ which it has gained from the moving space charge sheath. The collisions between electron and space charge boundaries at both electrodes send the electron into the plasma with energies capable of sustaining ionization. The contribution of the rf sheath's boundary movement, by the surfriding effect for sustaining the plasma at low excitation frequency, will be less since the sheath boundary velocity will be much smaller at lower frequencies.

This energy momentum transfer of an electron collision with a space charge boundary, where the electron appears to recoil from an apparent "solid" space charge boundary and thus contribute to ionization, is very difficult to prove experimentally and so far has eluded verification.

Microwave discharges do not appear to rely on secondary electron emission via ion bombardment and may have only very small electric fields present across the plasma to electrode surface. The existence of space charge sheaths at these very high frequencies (10^9 Hz) is questionable. For a given electric field to pressure ratio, (E/P), microwave discharges appear to be richer in free radical and excited species and this may be due to the radiation itself directly coupling into the molecules (124) (125). Plasma etchers using microwave excitation, are used as down stream reactors since they produce large numbers of free radicals. These are pumped away from the plasma volume and passed over the sample to be etched before being pumped away. However, etching in this type of system is isotropic because of the absence of ion bombardment.

The electron energising mechanisms to sustain ionization in plasmas may be investigated by measuring the electron temperatures (T_e) in the system, at various conditions of pressure and frequency. The electron temperature will change with pressure and must have a reasonable number of collisions in order to produce a maxwellian energy distribution. Using probes Polman (126) observed that electron temperature increased with decreasing pressure to a maximum at 20 mtorr pressure. He observed a violation from what was considered the maxwellian (type) energy distribution of electrons below 20 mtorr pressure. This effect was also observed by Vacquve et al.(127) and their results are shown in fig. 6.12.

Probe measurements in a planar reactor system at high pressures (300 mtorr) and low frequency (380 kHz) corresponding to a high ion density plasma have estimated electron temperatures of a probable maxwellian distribution to be as high as 16 eV for a CCl_4 plasma (130). This high electron temperature (T_e) result is perhaps due to the higher plasma potentials developed in the system. T_e will tend to increase with a decrease in pressure and is also affected by the ionization potential of the gas species in the plasma. In this system there was a reduction in operating voltage as the pressure increased from 100 mtorr to 500 mtorr (15).

In a plasma high energy electrons can ionize, excite and rupture molecular bonds and be lost rapidly by diffusion to the walls and then recombining. Thus we might expect to have fewer high energy electrons than a maxwellian distribution predicts (28). At pressures above 100 mtorr and at an excitation frequency below 1 MHz the plasma ionization conditions are significantly affected by the electrode surface material (129), since after the bombardment by energetic ions the surface emits secondary electrons. The surface efficiency for emitting secondary electrons depends on its secondary electron coefficient (γ), which has a small variation with ion energy and mass.

At low frequency, larger operating potentials are required.

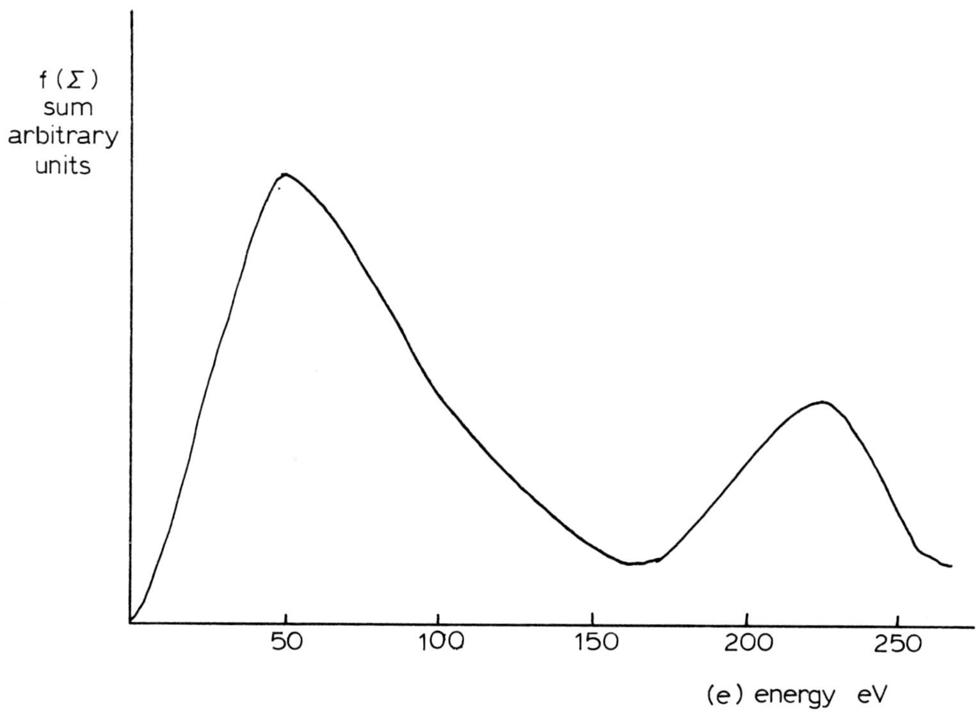

Fig 6.12

The electron energy distribution of a plasma system at pressure below 20 mtorr from ref 127

This will increase the field across the space charge sheath which will also increase in width (128). A secondary electron produced at the electrode surface will be accelerated by the space charge field and will have an energy determined by the sheath potential at the time of production. This electron will then carry out its statistical collision processes in the plasma. In addition, with low frequency plasmas, secondary electrons that are accelerated out into the plasma will have initial energies related to the variation of rf potential. This variation in sheath potential will produce a wide spread of energies. The secondary electrons that are produced at the rf peak value will have considerable energy.

If we now consider other changes that can affect electron distribution, i.e. reducing the operating pressure, there will probably be an increase in applied potential to the rf electrode. This will also coincide with an increase in the dc negative bias to the rf electrode. The increase in bias occurs because the plasma diffusion path has increased and changed the symmetry of the system. In this case it is possible that the rf electrode has developed a much larger space charge sheath electric field than at the gound electrode. Thus secondary electrons that are accelerated via the sheath and emitted into a plasma from these electrodes will have two energy distributions, probably similar to fig. 6.12.

6.4 Summary of Sustaining Mechanisms

1. Ions are accelerated through the space charge sheath impacting onto the electrode surface producing secondary electrons which are accelerated out into the plasma by the space charge sheath to continue the ionization collision processes.

2. Secondary electrons emitted from the electrode surface ionizing neutrals within the dark space producing more ions and thus an avelanche or multiplication condition occurs.

3. Electrons which impact onto a surface with large energies and when only opposed by a small space charge field on arrival, can produce secondary electrons. These secondary electrons are then accelerated out into the plasma by the now increasing space charge field.

4. The energy transfer to the plasma electrons, may arise from piston or surfriding motion of the space charge sheath boundary.

5. The energy transfer to the plasma electrons may be achieved by collisions with fast neutral atoms or molecules in the plasma.

6. Photo ionization:- the production of secondary electrons by far UV photons generated by metastable excited species and recombination events in the plasma. The photon arriving at the surface can give a photo-electric effect, but in plasmas the UV generation is probably small.

Probe measurements in low frequency planar systems 300 mtorr pressure have revealed electron temperatures of 4 eV for O_2 and Ar, and 16 eV for CCl_4. The high electron temperature is probably due to the increase in operating voltage observed for the CCl_4 plasma compared with O_2 or Ar at the same power and pressure. This effect is probably due to the extra energy required to cleave and ionize the CCl_4 molecule and the larger recombination events in the CCl_4 plasma (130).

The recombination mechanisms which may be high for the large cross section of CCl_3^+ with slow electrons, there may also be a possibility of Cl^- recombination with CCl_3^+. The effect of this recombination will be to reduce the charge density in the plasma. If the system runs at constant power, then to maintain the ion density the operating voltage must increase. Formation of Cl^- is likely with slow electrons since

$Cl^{\cdot}$ has a high electron affinity:

$$Cl^{\cdot} + e \rightarrow Cl^{-} + 1.12 \text{ eV } (\Delta H)$$

6.5 System Voltage Distribution

In a symmetrical system where the electrode areas in contact with the plasmas are equal then the voltage drop across the dark spaces is the same. Even the apparent symmetrical systems (15) show some difference in sheath potentials by developing a dc negative bias to ground on the live rf electrode, due to the difference in area/ion collection of the sheaths.

For sputtering applications the system would not be successful if the dark space sheaths were equal since material would deposit and redeposit from each electrode. Sputtering systems were designed to be asymmetrical so deposition occurred on one surface only. Koenig and Maissel (108), considering this, formulated an expression between voltages and electrode areas. The assumptions were a) No collisions of ions traversing the dark space, thus Child's space charge limited current expression can be valid.

$$J_i = \frac{KV^{3/2}}{Mi^{\frac{1}{2}}D^2}$$

where:-

J_i = current density

K = permittivity constant

M_i = mass of ion

D = dark space width

V = Voltage across dark space

b. The current density of the ions M is uniform and equal at both electrodes, thus:-

$$\frac{V_1^{3/2}}{D_1^2} = \frac{V_2^{3/2}}{D_2^2} .$$

c. The capacitance (C) across the charge sheath is proportional to the electrode area (A) and inversely proportional to the dark space thickness

$$C \propto \frac{A}{D} .$$

d. Combining C and d and then substituting b

$$\frac{V_1}{V_2} = \frac{A_2}{D_2} \cdot \frac{D_1}{A_1} = \frac{V_1}{V_2} = \left(\frac{A_2}{A_1}\right)^4 .$$

Koenig and Maissel's work was at a few mtorr and 13.5 MHz operating frequency. The fourth power term in real systems tended to enlarge the effect of geometry (108). The actual areas may not be measurable since the area of contact between the electrode and the plasma is difficult to estimate.

Changes in pressure will affect the diffusion rate of the ions and change the effective areas in contact with the plasma as well as the voltages. Jackson (5) has shown the reduction in dc bias as the pressure increases in a rf sputtering system.

Coburn and Kay (32) have measured the relationship between target and substrate sheath voltages in an rf diode system; (14.5 MHz) as a function of electrode area ratio. This was achieved by using a series of pyrex confining cylinders all of equal height but ranging in diameter. Figs. 6.13 and 6.14 show the apparatus used and the plots of substrate sheath voltage against target sheath voltage for various area ratios.

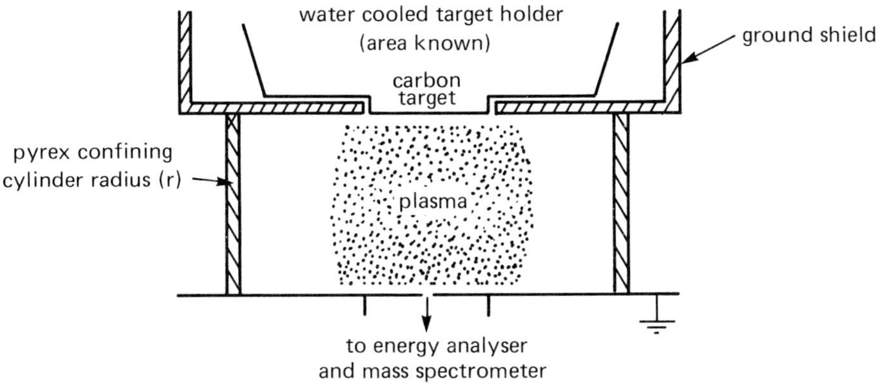

Fig. 6. 13 Apparatus used by Coburn and Kay ref 32

The plasma potential in Coburn and Kay's system was estimated from the measurement made with a quadrupole mass spectrometer fitted with an energy analyser. This system measured the ion energy from plasma to ground and could also identify the type of ion species.

The Koenig and Maissel model predicts that the ratio of substrate to target voltage should be constant for a given area ratio and this appears to agree with Coburn and Kay's work for the fixed family of lines which appear to have a common intercept which is not at the origin. The Koenig and Maissel model also predicts that the sheath voltage ratio should be proportional to the fourth power of the inverse area ratio. However, values obtained by Coburn and Kay suggest that the power dependence should be much closer to unity.

Low frequency plasmas produce larger operating potentials compared with the same power density at higher frequencies. However high excitation frequency generates relatively much larger dc negative biases on the rf powered electrode and for this reason high frequency excitation systems are used for sputtering applications. The large dc

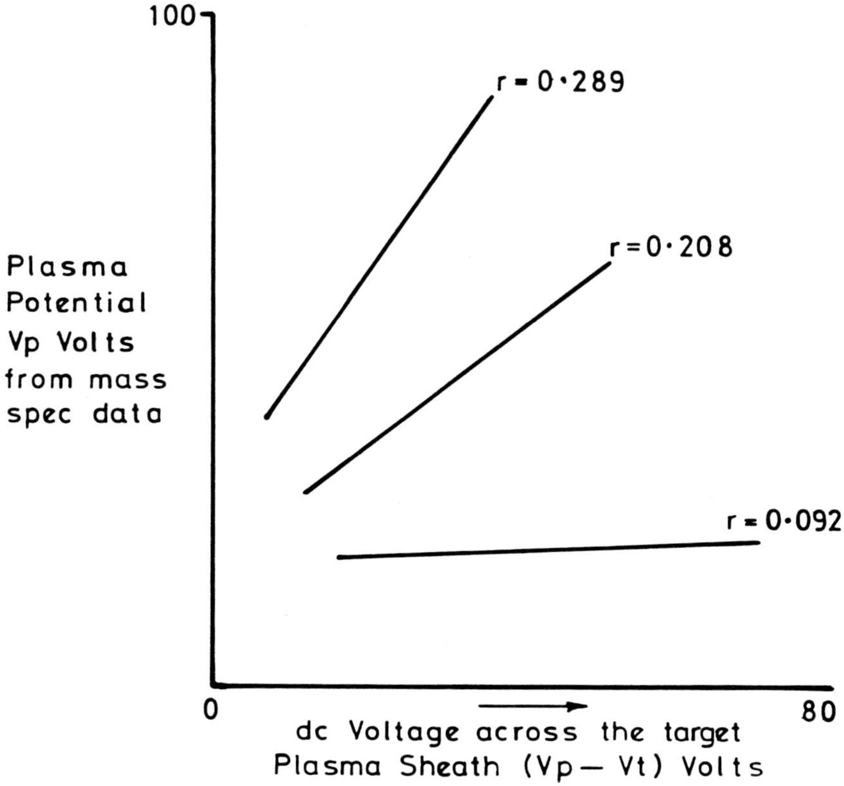

FIG 6.14

PLOT OF PLASMA POTENTIAL (ESTIMATED) AS FUNCTION OF DC VOLTAGE ACROSS TARGET TO PLASMA SHEATH AT DIFFERENT GROUND TO TARGET AREA RATIOS (32)

negative bias produced at the high frequencies is probably due to transient effects, where a large dc negative bias is needed so a small change in plasma potential will allow the electrons to surmount the dark space barrier. However, large dc negative biases to ground can still be produced at low excitation frequency by using large ground surface areas, but the ratio of applied potential to dc negative bias may still not be as large as in high frequency plasmas (67). This is because the plasma potential at high excitation frequency may only be a few volts above ground potential whereas the dc negative bias may be two orders of magnitude higher. Bruce (131), citing ref. 17, claims that large self-biasing occurs only at low pressures, especially with electro-negative gas such as Cl_2. This is not surprising since at low pressures the ion diffusion increases thus increasing the likelihood of asymmetry. By using an electro-negative gas such as Cl_2 the likelihood of recombination at these surfaces will increase in turn raising the operating potential and the dc negative bias.

CHAPTER SEVEN

PROBE CHARACTERISTICS AND PLASMA MEASUREMENTS OF AN ELECTROTECH PLANAR PLASMA ETCHER

7.1 Introduction

It was explained in previous chapters that rf flat bed reactors or planar reactors have become commonly used for plasma etching of semiconductor wafers used in device fabrication. One of several reasons for their adoption is that they are capable of producing etched lines with edges normal to the surface, thereby providing line width fidelity on very small geometries.

Previous discussions have shown that much has been done to determine the nature of the chemical species and reactions which occur in rf discharges with planar electrodes. There is, however, little information on the ionization conditions in typical production systems as used in device fabrication work in industry. During the research discussed here a production type flat bed planar reactor was used to investigate the plasma conditions that can typically arise when etching aluminium and polysilicon films which are used in NMOS and CMOS device fabrication.

This chapter reports on the measurements of electron temperature T_e and ion density N_i when using double probes and the work done in an attempt to determine the plasma potential V_p from single probe measurements. Also the problems associated with obtaining probe characteristics, when inert and active gases are present in an rf discharge, are considered.

7.2 Background

Before discussing the experimental work undertaken, a brief outline of the characteristics of probes used to examine plasma conditions is presented.

Probe diagnostics were first used by Langmuir and Mott-Smith in 1924 (ref. 134) and are the oldest and probably the simplest techniques for measuring the properties of a plasma. Unfortunately probe theory is complex and restricted in the conditions under which a particular theory can be applied. Extensive reviews can be found in the literature (135,136) covering many experimental and theoretical variations.

The current flowing to a probe biased with respect to a reference electrode or second probe is measured as a function of the applied bias voltage, to give an I/V characteristic. The I/V characteristic obtained depends on the plasma conditions, site and position of probe, on the geometry and radius of the probes and reference electrode used. Probes usually have shapes such as spheres, cylinders and planes whose areas can easily be defined.

Probes can also be classified by their dimensions relative to the mean free path of the charged particles in the plasma and to the Debye screening length (λ_D) in the plasma. Six assumptions or requirements are generally made when using probes in plasmas, they are:

(i) The plasma investigated tends to be infinite in volume and is homogeneous and quasineutral (i.e. has charge balance) in the absence of the probe.

(ii) Electrons have a Maxwellian energy distribution with the electron temperature considerably greater than the ion temperature $T_e >> T_i$.

(iii) The mean free path of electrons and ions is large in comparison with any space charge sheath around the probe so that particle collisions do not occur in the sheath. For this reason probes are more successful for investigating plasmas operated at low gas pressure, i.e. below 100 mtorr,(135).

(iv) The particles colliding with the surface of the probe should not sputter atoms, eject electrons or react with the probe material. Thus probes are usually made of materials which are chemically inert, have a high working temperature, high work function, and high sputtering threshold. Probes are more successful in low ion density plasmas, i.e. below $10^{17}\,m^{-3}$.

(v) The probe size is usually as small as possible to minimise perturbation of the plasma. This condition can usually be met by the probe current, i.e. rate of charge loss from the plasma, being less than 5% of the total rate of charge flow into the plasma.

(vi) When using a single probe, the area of the reference electrode, including the chamber wall, should be sufficiently large to collect a positive ion current of sufficient magnitude so as not to limit the electron current flow to the probe before it reaches the plasma potential. Thus the reference electrode, e.g. a grounded work holder and chamber walls, should have good contact with the plasma to ensure the ideal conditions for measuring the I/V characteristic are approached.

7.2.1 Probe theory and the single probe

There are two fundamental conditions on which probe theory is based. The first is where the probe radius r_p is many times larger than the Debye screening length (λ_D); $r_p >> \lambda_D$. Under this condition a limiting case solution is obtained where all charged particles that enter the probe sheath are collected. The electron current to the probe i_e is then given by the relation:

$$\ln i_e = \ln i_{es} - \frac{eV_s}{kT_e} \tag{1}$$

The symbols used in this section and those in equation (1) are:

V_s	is the voltage across the ion space charge sheath surrounding the probe, i.e. $V_s = V_p - V_B$
V_p	is the plasma potential with respect to ground (reference electrode)
V_B	is the applied voltage between the probe and the ground reference electrode
V_f	is the floating potential, i.e. the voltage with respect to a reference electrode at which the total probe current is zero
e	is electron charge
k	is Boltzman's constant
T_e	is the electron temperature
i_{es}	is the probe electron saturation current
i_{is}	is the probe positive ion saturation current.

In practice, one cannot apply a potential between a probe and the plasma, but must use a reference electrode. If the latter is a large area plate immersed in the plasma, then at zero applied voltage both the reference plate and probe will float at the same potential to the plasma so that the random electron and ion current flows to their surface are balanced and there is no external current flow. The floating potential of an immersed body with respect to another electrode is designated as V_f (the other electrode is usually the ground electrode).

If a potential is applied to a probe with respect to a reference plate which also serves as an excitation electrode in a dc or r.f. discharge system, then the reference plate may be more biased to the plasma than a floating body. This is the situation for the probe characteristic illustrated in Fig. 7.0 where the reference electrode has a negative potential to the plasma potential greater than that of a floating body. The thick curve in Fig. 7.0 is an I/V probe characteristic for the condition $r_p \gg \lambda_D$.

In this curve the region A-B is referred to as the random positive ion density or ion saturation region i_{is} and is where ion collection by the probe alone occurs at large values of negative potentials on the probe with respect to the reference electrode. As the probe becomes less negative with respect to the plasma (i.e. V_s becomes smaller) some of the higher energy electrons may reach the probe. This occurs providing they have sufficient energy to overcome the potential V_s across the probe sheath in the region B-C. Thus when the floating potential is reached,the random electron and ion charge flow are balanced and the probe current is zero. Some electron retardation occurs until finally the probe reaches the plasma potential at point 'D' so that the probe acts as an energy selector.

The electron probe current i_e is easily estimated by addition of the measured probe current and the ion saturation current i_{is},which also flows in that region. Increase of positive ion current with negative bias arises from the external area of the ion sheath growing and increasing the ion current collected. In practice, the value of i_{is} (positive

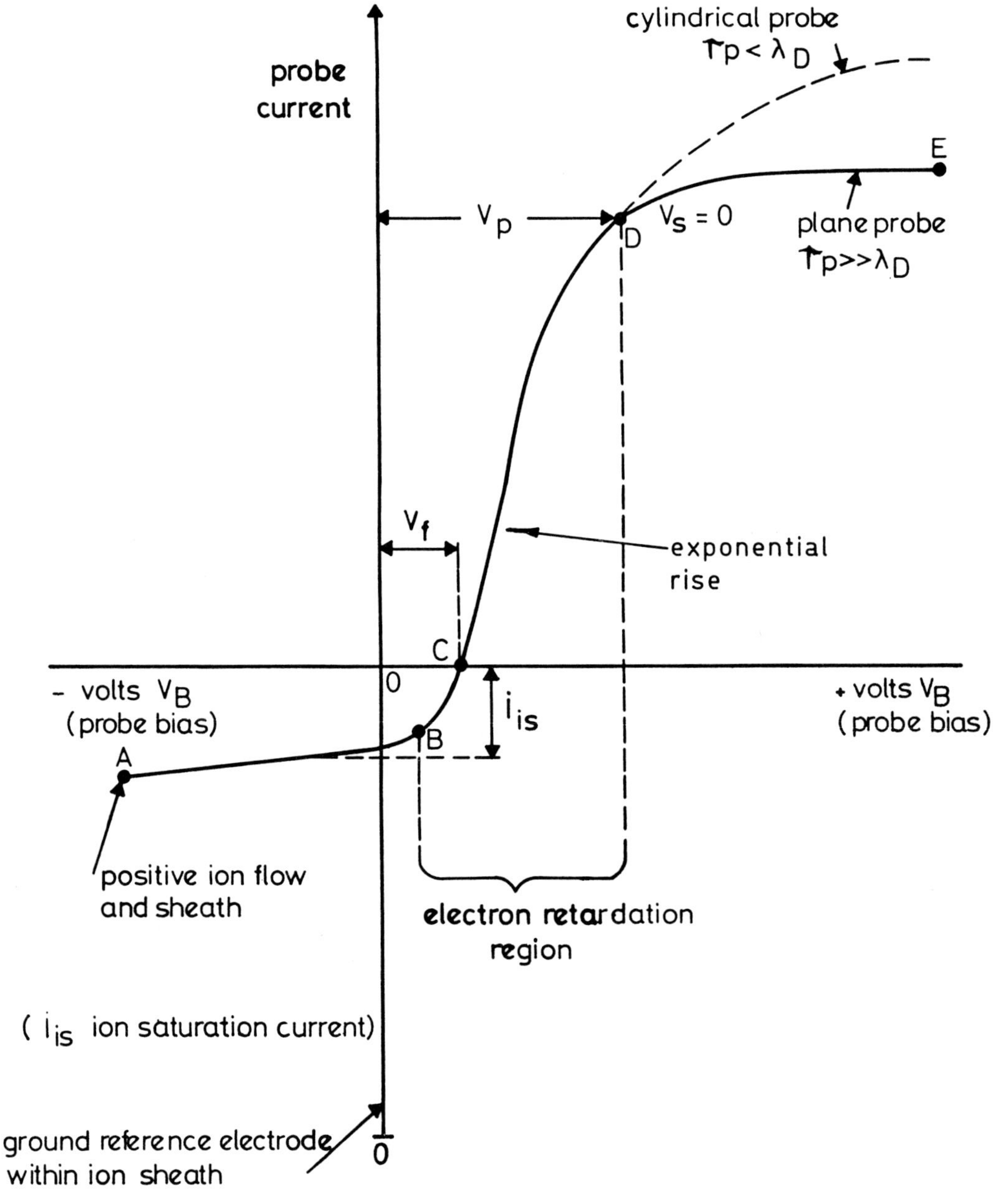

FIG 7.0

CURRENT - VOLTAGE CHARACTERISTIC FOR A SINGLE PROBE USING THE PLASMA VESSEL AS THE REFERENCE ELECTRODE.

ion current) is small because of the lower ion temperatures, and thus may be neglected for single probe work.

As the probe potential approaches that of the plasma potential ($V_s \rightarrow 0$) the electron flow rises until it reaches a saturation value i_{es} given by the random rate of arrival of the electrons at $V_s = 0$. Thus a plot of ln i_e against probe voltage in the region B-D should give a straight line of slope $e/k\,T_e$ from which T_e the electron temperature can be derived.

At the point D (where $V_s = 0$) there is no sheath around the probe and thus the current collected is at its maximum saturation value i_{es}. Knowing T_e, the probe collection area A and the random electron current density J_{es} the electron density N_e can be obtained from

$$J_{es} = N_e\, e(\frac{\bar{v}}{4}) \tag{2}$$

where the average electron velocity

$$\bar{v} = \sqrt{\frac{8kT_e}{\pi\, m_e}} \tag{3}$$

and m_e is the electron mass.

As, $\frac{i_{es}}{A} = J_{es}$ then N_e is given by

$$N_e = \frac{4i_{es}}{A\, e\, \bar{v}} \tag{4}$$

The other probe condition is where the probe radius r_p is less than λ_D. In this case the charged particle collection is known as orbital-motion limited and the current collection and flow has been derived in ref (137). The probe current is given by,

$$(i_e)^2 = \frac{4}{\pi}\,(i_{es})^2\left(\frac{eV_s}{kT_e} + 1\right) \tag{5}$$

This relation can be expected to hold for very small cylindrical probes with thick space sheaths. Under such conditions a plot of $(i_e)^2$ against V_B should give a straight line of slope S which can be expressed by the equation

$$S = \frac{4(i_{es})^2 e}{\pi k T_e} \qquad (6)$$

An I/V characteristic probe plot using a small probe for orbital motion collection ($r_p < \lambda_D$) is shown by the broken curve in Fig. 7.0, illustrating a poor change of slope at point D. An $(i_e)^2$ versus voltage (V_B) plot for a specific system using a small single probe is shown in Fig. 7.17. The line from this plot cuts the voltage axis at V_C; T_e can be obtained because:-

$$T_e = \frac{eV_C}{k} \qquad (7)$$

The plot shown in Fig. 7.17 is discussed more fully further on.

7.2.2 Double probe

A later development of the Langmuir probe principle uses two identical probes biased solely with respect to each other (138). Thus, unlike the large reference electrode previously considered, the electron current flowing to the positively biased probe is limited by the positive ion current flowing to the other. A typical experimental arrangement and probe characteristic is shown in Fig. 7.1. The I/V curve is symmetrical for the negative and positive values of current. The saturation for electron flow arises from ion current limitation and not because the plasma potential is reached. The double probe system will disturb a plasma less than a single probe arrangement since the maximum current drawn is limited by the maximum ion current collected at saturation (i_{is}) by the probe. However, the double probe system cannot

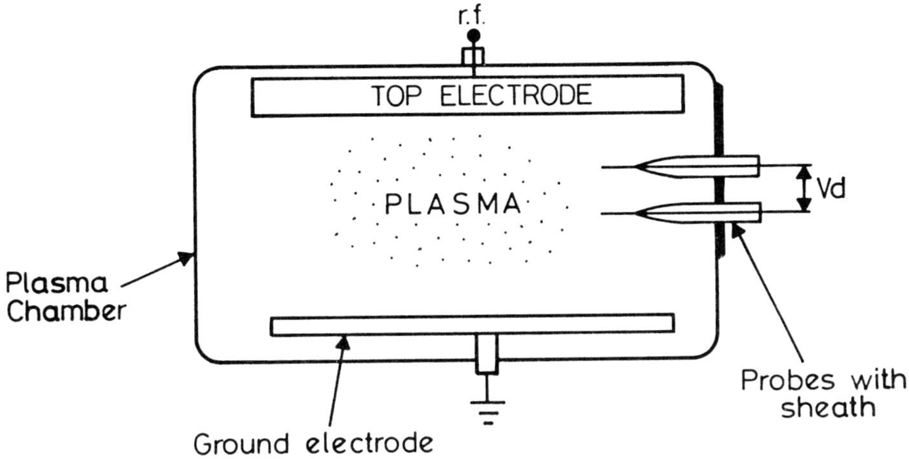

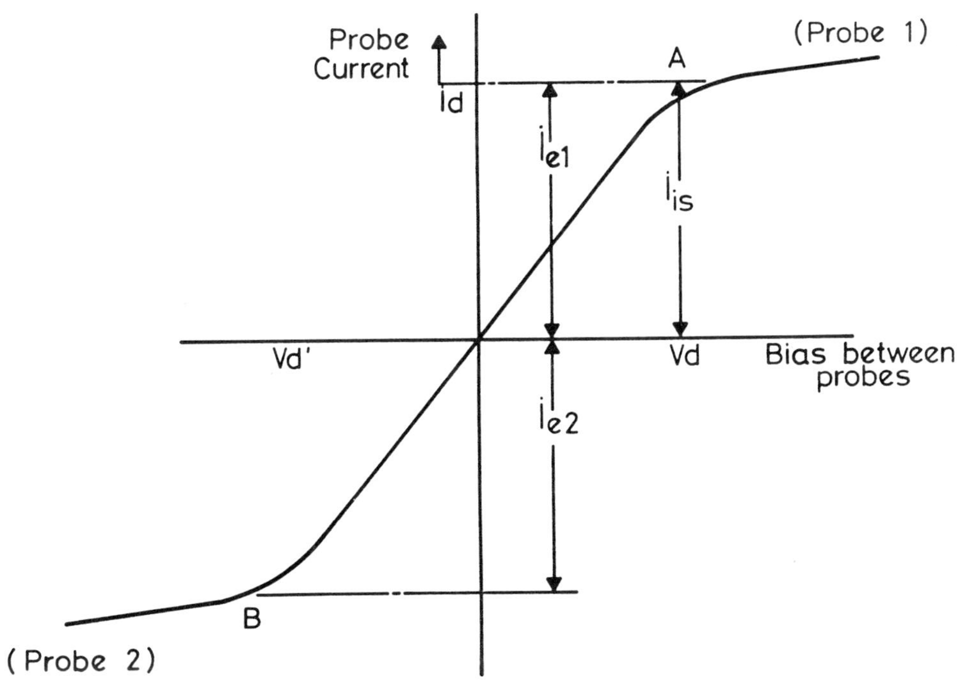

FIG 7.1

THE DOUBLE PROBE CHARACTERISTIC & ARRANGEMENT.

directly give the floating potential of a body with respect to a probe at plasma potential.

Both plasma potential and floating potential measurement can be important when determining ion impact at a process surface. Although double probes cannot measure plasma potential or floating potential they can give electron temperature and charge density values. The ion impact energy on process surfaces is discussed in the next chapter.

When the probes are floating, and there is no applied potential between them, the probes develop the same negative bias to the plasma as the electron flow is limited to balance the positive flow of the ions. It may be assumed that both probes collect the random ion current. For symmetrical probes of equal area (A) with zero applied potential between them, the electron and ion currents are equal for both probes since the probes have similar sheaths and act exactly like separate floating probes.

If one probe, termed 1, is now made negative to the other (probe 2) by applying a potential between them, then the probe to plasma potential of probe 1 (V_1) is increased. Thus the electron current of probe 1 (i_{e1}) decreases. Similarly, at probe 2, V_2 decreases and i_{e2} increases. The excess current flows in the external circuit between the probes as a measured current i_d.

Probes with respect to plasma potential, V_1 and V_2, are related to the applied potential (V_d) as follows:

$$V_d = (-V_1) + (-V_2) \qquad (8)$$

Both probes remain negative to the plasma as neither can reach plasma potential. Although probe 2 is nearer to the plasma potential, it cannot be reached because the maximum external current is limited by the positive ion current to probe 1 when $i_{e1} = 0$ ($i_{e2} = i_d + i_{is}$).

The voltages $(-V_1)$ and $(-V_2)$ assume values which not only make their difference V_d but which retain the total electron current equal to the total ion current $2i_{is}$ which is constant (since $i_{e1} + i_{e2} = 2i_{is}$).

The measured current i_d at saturation is 1/2 of the difference between the electron currents and is equal to i_{is}, i.e.

$$2i_d = i_{e2} - i_{e1} \tag{9}$$

When V_d is sufficiently negative for probe 1 to prevent electron arrival, probe 2 collects excess electrons to balance in the external circuit the ion current to probe 1, then the saturation point A is reached. This is shown in Fig. 7.1. Any further increase in V_d drives probe 1 more negative to the plasma without, under ideal conditions, moving probe 2 near to the plasma potential. As the potential drop across the ion sheath rises to the saturation current level the sheath can change in diameter so that the current does not reach a constant value with voltage.

As the I/V characteristic is symmetrical point B is similar to that of A. Considering the region AB, which is the unsaturated region, this can be regarded as the performance of two single probes which follow the relations:

(probe 1) $$i_{e1} = i_{es}\, e^{-\frac{eV_1}{kT_e}} \tag{10}$$

(probe 2) $$i_{e2} = i_{es}\, e^{-\frac{eV_2}{kT_e}} \tag{11}$$

As $i_{e2} + i_{e1} = 2i_{is}$ and combining (10) and (11) thus

$$2i_{is} = i_{es}\left(e^{-\frac{eV_1}{kT_e}} + e^{-\frac{eV_2}{kT_e}} \right) \tag{12}$$

The calculation of double probe characteristics by the semi-log method is well described in Johnson and Malter (ref. 138). An error arises in the semi-log method because of difficulty in determining the probe ion current saturation value (i_{is}) i.e. when there is no sharply defined region. Such an effect is commented on in this chapter for a carbon tetrachloride plasma. However, in this method the error may be reduced since any error in the selected value of i_{is} introduces an error in the same direction in the value of i_{e2} thus the value Γ tends to be less affected.

Semi-log plots involve some laborious computation of $\ln\Gamma$ for each voltage value, but electron temperatures can be determined considerably more easily by using the 'equivalent-resistance method' (135). Dividing by i_{e2} then:

$$\frac{2i_{is}}{i_{e2}} = e^{-e\frac{(V_1-V_2)}{kT_e}} + 1 \tag{13}$$

$\frac{2i_{is}}{i_{e2}}$ can be easily estimated from the I/V characteristic shown in Fig. 7.1 and by simplifying equation (13) using $\Gamma = \frac{2i_{is}}{i_{e2}} - 1$.

Thus*

$$\ln\Gamma = \frac{e}{kT_e} V_d \tag{14}$$

Γ, can be regarded as

$$\Gamma = 2 \times \frac{\text{ion saturation current at probe 1 or 2}}{\text{current drawn } i_e\text{, for probe 2 at } V_d \text{ volts}} - 1$$

A graph of $\ln\Gamma$ against V_d (probe applied potential) yields a straight line of slope $-e/kT_e$ from which T_e can be calculated. A typical plot with the method of estimating T_e is shown in Fig. 7.3 and is called the 'semi-log method'.

---

* $(V_1-V_2) = V_d$.

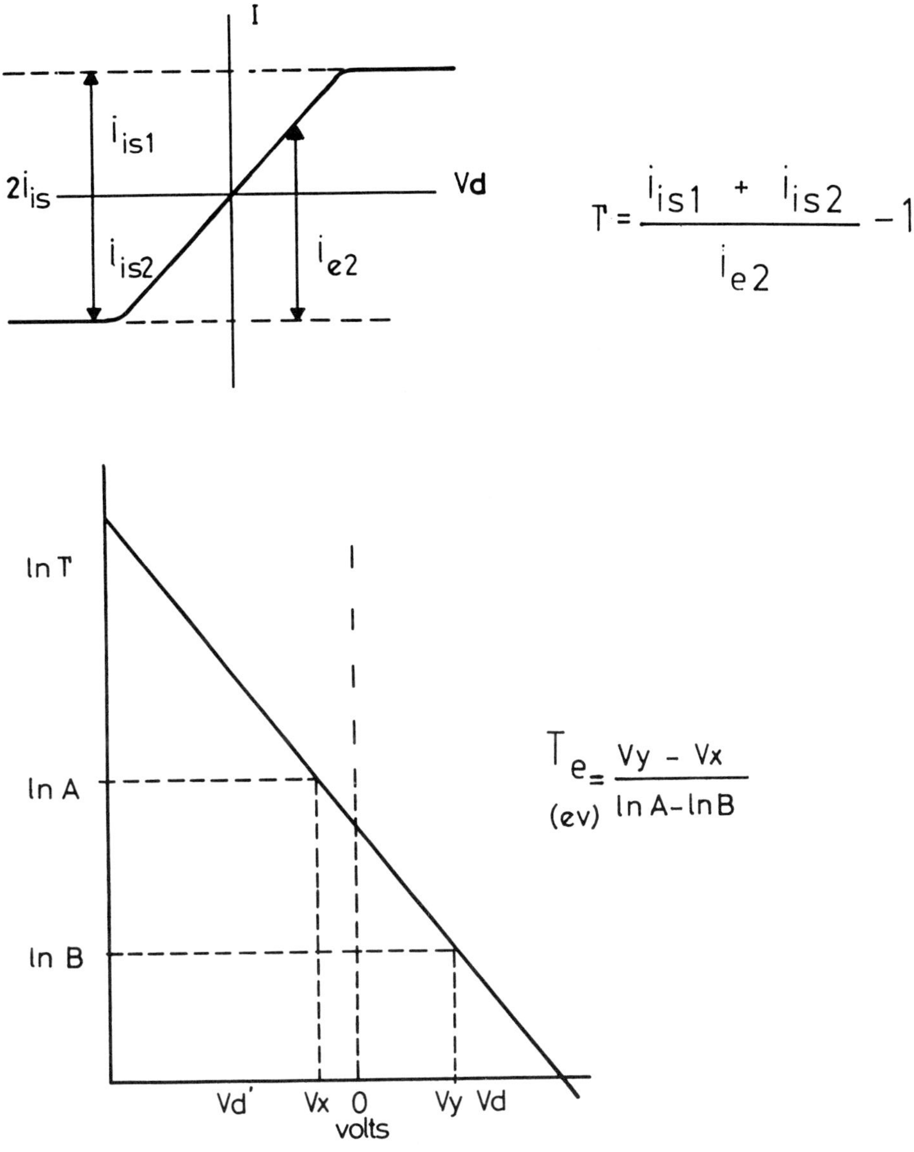

FIG 7.3

SEMI LOG METHOD USED FOR ESTIMATING THE Te OF A PLASMA FROM PROBE PLOT.

This is a simple method, generally favoured for symmetrical probes where the equivalent resistance R_o^{-1} is measured at the slope value where $V_d = 0$, from which T_e can be estimated from:

$$T_e = \frac{e}{4k} 2i_{is} R_o$$

(i_{is} probe ion saturation value).

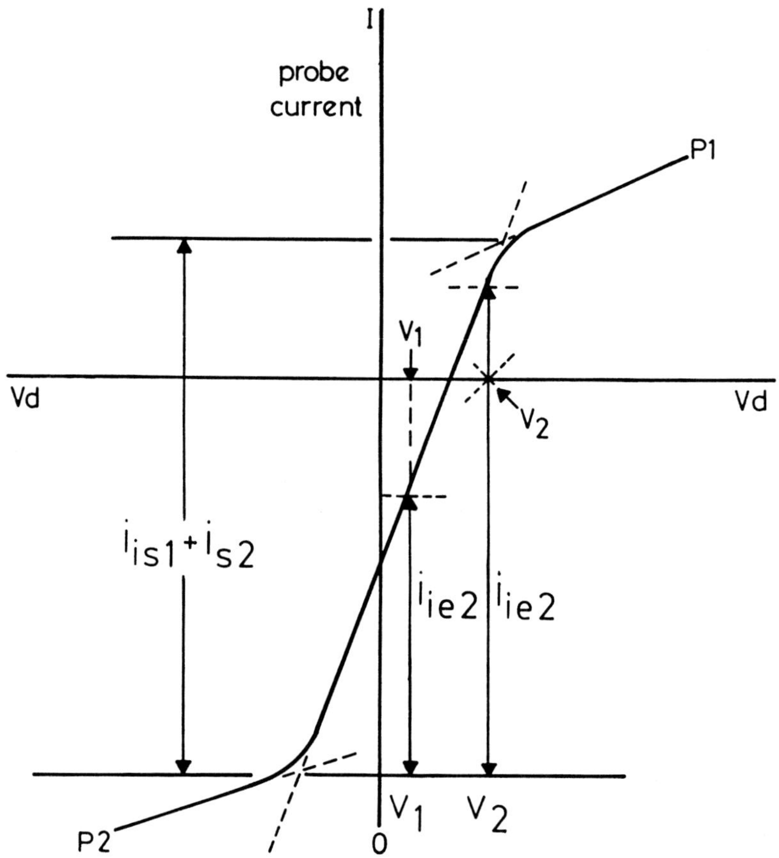

FIG 7.4

T_e CALCULATION BY THE INTERCEPT METHOD

If the probes are asymmetric the I/V characteristic is unsuitable for this method (i.e. no $V_d = 0$), but in this case a rapid method of calculating T_e can be used, known as the 'intercept method' as shown in Fig. 7.4. In the 'intercept method' one uses the relations:

$$D = \frac{i_{is}(\text{Probe 1})\, i_{is}(\text{Probe 2})}{i_{e2'}}$$

with i_{e2} measured at V_2 and

$$F = \frac{i_{is}(\text{Probe 1}) + i_{is}(\text{Probe 2})}{i_{e2'}}$$

with i_{e2} at V_1. From these one can derive the T_e from

$$T_e = \frac{V_2 - V_1}{\ln\left(\frac{F-1}{D-1}\right)}$$

All three methods will be used here to calculate T_e values from the probe measurements in a planar reactor.

7.2.3 Limitations

For device fabrication using a planar plasma etcher, the optimum conditions for etching 1 µm aluminium 2% silicon films were observed to be at 300 mtorr pressure using a CCl_4 plasma with a current input of 2 Amps. These conditions gave an etch rate of 3000Å per min with a selectivity of 10:1 for Al/2% Si-films on a silicon dioxide substrate. The selectivity between aluminium and the masking photoresist layer was approximately 2:1.

The criteria laid down for probes in section 7.2 may not be met under all the conditions which can exist in the etcher. For example, the electron mean free path in CCl_4 plasma at 300 mtorr pressure has been estimated to be about 0.35 mm and this may be smaller than the probe space charge sheath depending on the ion density and temperature. If electrons collide in the probe sheath it is difficult to assess their effect because it is now in the complex collision dominated regime (148). It is believed that such an effect should not prevent measurement of floating potential, plasma potential and estimation of ion current. Also as shown later, more serious problems were encountered from the effect of high density plasmas on immersed probes such as temperature rise.

7.3 Experimental Work and Results

7.3.1 Apparatus

The system used was an Electrotech 317 reactor, which had two water cooled planar electrodes fabricated from anodised aluminium. The electrodes were 60 cm in diameter with a separation of 5 cm. A simplified diagram of the apparatus is shown in Fig. 7.5.

A disk of anodised aluminium perforated with 17 holes rested on the grounded lower electrode. A small disk of anodized 0.8 cm thick and 7.6 cm in diameter was located in each hole; these disks were used to support the wafers during processing. Gas was admitted through the top of the chamber via the top electrode to the reaction zone. Waste gas was pumped away by a rotary pump (Alcatel 2030) connected to the base of the system via a gate valve (Fig. 7.5). The pressure in the system was measured with a Baratron diaphragm capacitance gauge which was located between the chamber and the gate valve. The power supply was a 6 KW 380 kHz r.f. generator and this will be described in more detail in the next chapter. The conditions investigated by probes were between 100-500 mtorr pressure and rf currents of between 0.5 to 5 Amps rms, using carbon tetrachloride and argon plasmas.

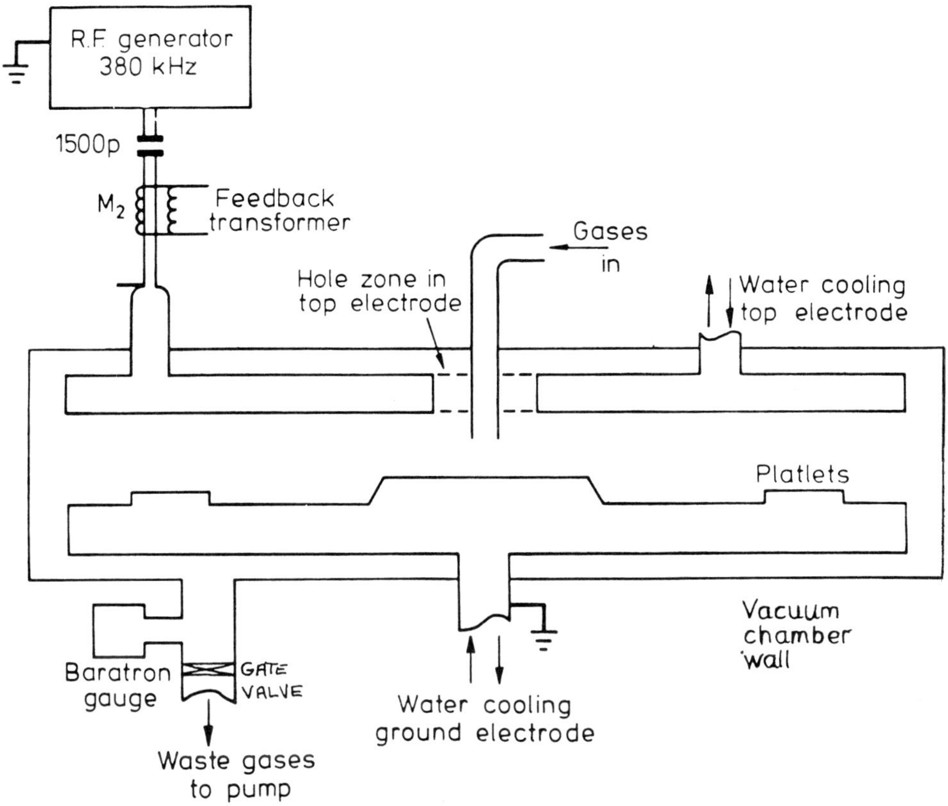

Fig 7.5

The Basic 317 Planar Reactor

7.3.2 Single probe experiments

The experimental arrangement used to measure the plasma characteristics is shown in Fig. 7.6. In all the experiments, co-axial cable was used throughout to reduce rf interference. A choke (1 mH), was connected to the probe to prevent the r.f. component of the collected current from entering the measuring circuit. A variable potentiometer, connected to batteries and a d.c. voltmeter (screened in a box) was used to bias the

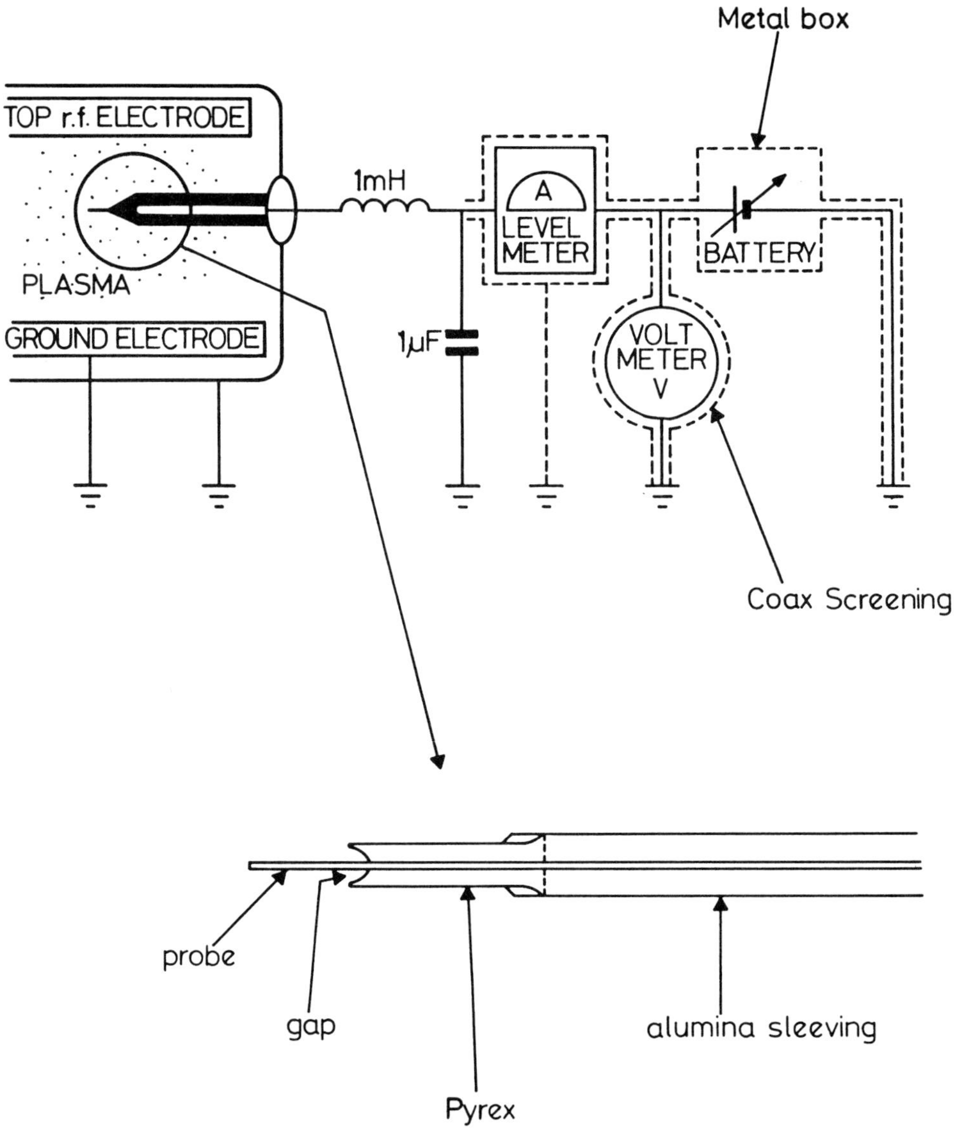

FIG 7.6

EXPERIMENTAL ARRANGEMENT INCORPORATING PROBE FOR PLASMA DIAGNOSTICS.

probe from -120 to +1200V if needed. The probe current was measured using an electrometer (Level TM9B).

The probes used first in the experiments were cylindrical 50 μm diameter Pt/Rh-wire and then a cylindrical 0.12 mm diameter rhodium wire was used. Both types of probe were coated and insulated with Pyrex glass and sealed at the end to leave a 4 mm length of wire protruding The probes were sheathed individually in alumina sleeving for support. The 4 mm length of probe was left protruding as shown in Fig. 7.6 to reduce any problems caused by spherical end effects collecting charge which would affect the probe characteristic. Also a gap was left between the probe and sleeve to limit the possibility of electrical shorts caused by the deposition of any conductive sputtered material. This is shown in Fig. 7.6.

The probes used in the experiment were mounted on a metal flange which also housed the electrical feed throughs. This flange was bolted to a port at the side of the vessel previously occupied by the observation window. The flange and the chamber were grounded and acted as the reference electrode when a single probe was used. Probes were placed midway between the electrodes of the system and penetrated about 20 cm from the wall into the chamber.

Cleaning of a probe is essential to obtain a reproducible characteristic and when oxygen plasmas were investigated with a single probe it appeared to 'age' before a reproducible characteristic could be obtained. This may be due to surface oxide formation which would change its characteristics.

Before use probes were precleaned in alcohol and trichloroethane solvents and then treated for 20 min in an argon plasma <u>in situ</u> (500 mtorr pressure,2 Amps rms) with a negative bias of -10 volts to clean their surfaces. This was found to be necessary to obtain reproducible results, because contamination could change the probe characteristic (ref. 136). Once the gas under investigation was introduced, a period of 20 min was allowed for the system to stabilise before any measurements were made. During this period the chamber was preconditioned as traces of material from previous experiments are removed.

When a single probe was operated at small positive potentials large currents were drawn and it began to melt. Each single probe used gave an exponential rise in current at low positive voltage bias and then the characteristic changed to a linear relation before finally melting.

Fig. 7.7 shows single probe plots for 50 μm diameter Pt/Rh-wire probe in an Ar-plasma at 100, 300 and 500 mtorr gas pressure with a plasma r.f. current of 2 Amps rms. The applied r.f. potentials measured in rms at these conditions were 240, 180 and 150 volts rms for each pressure respectively. The measurement of r.f. potential is discussed in the next chapter.

The probe voltage bias range was from +15 volts to -15 volts with the grounded components acting as a reference electrode. Provided the bias voltage range did not exceed this small positive value the probe I/V plots were reasonably reproducible. However, the melting did not permit determination of the plasma potential from the probe characteristic particularly as plasma potentials to ground were expected to be several hundred volts positive, i.e. a d.c. potential at maximum could be equivalent to the r.f. peak value. This aspect is discussed further in Chapter 8.

FIG 7.7

I/V SINGLE PROBE PLOTS USING 50 μm DIA Pt/Rh PROBE IN AN Ar PLASMA AT 2amps RMS 100,300 AND 500mtorr PRESSURE.

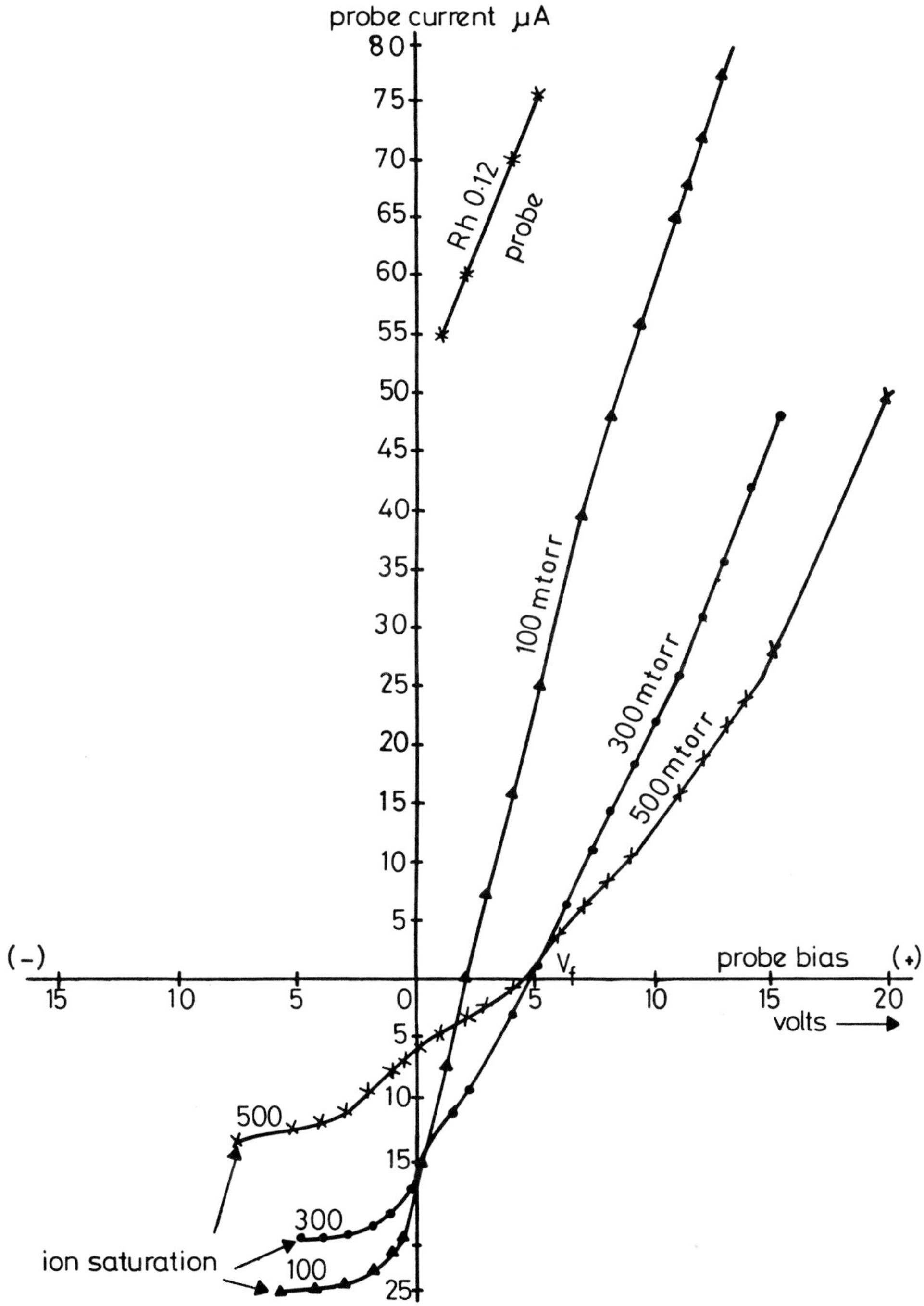
probe current μA
80
75
70
65
60
55
50
45
40
35
30
25
20
15
10
5
Rh 0.12
probe
100 mtorr
300 mtorr
500 mtorr
(-)
probe bias
(+)
15
10
5
0
5
10
15
20
volts
V_f
500
300
100
ion saturation

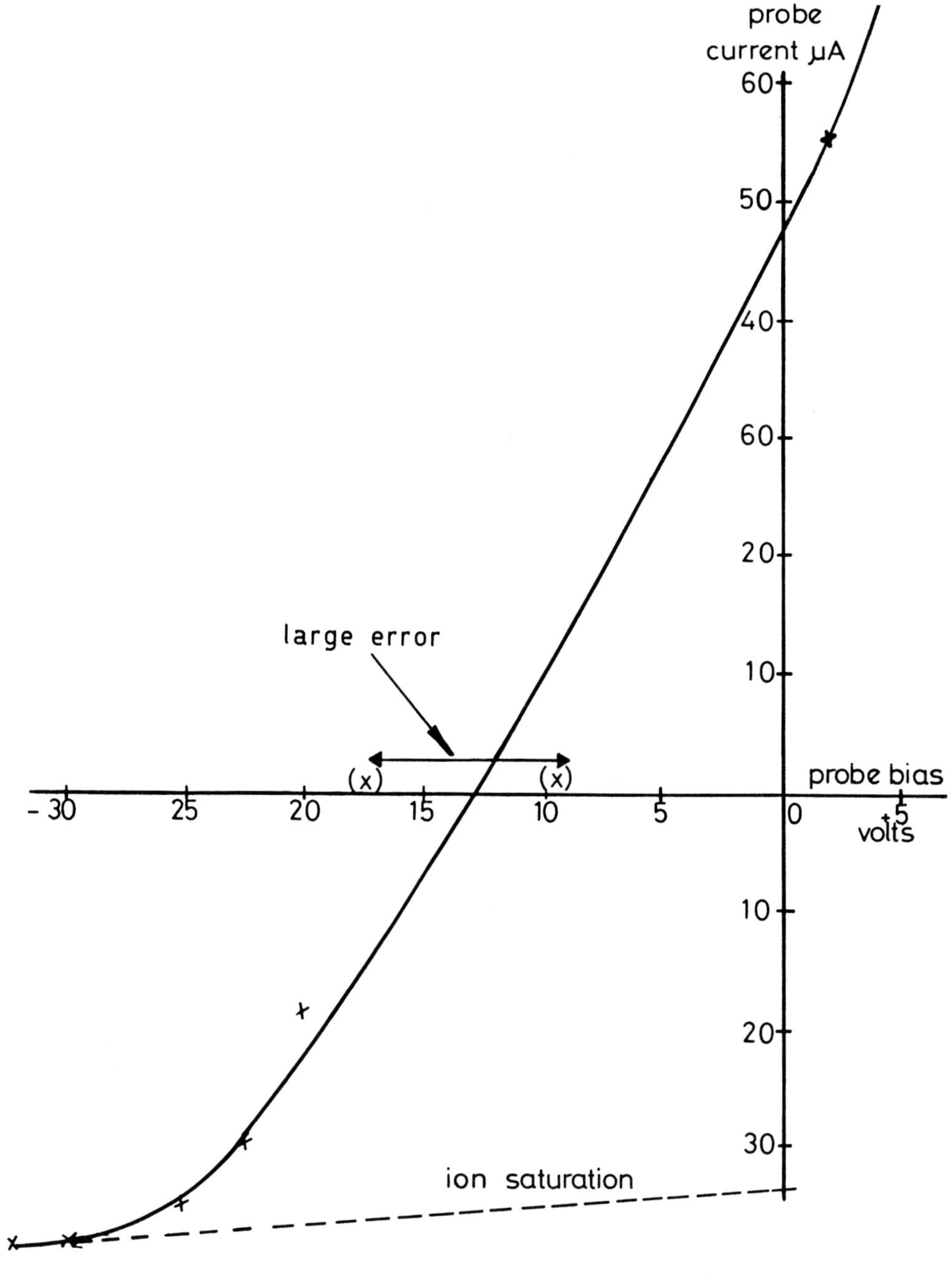

FIG 7.8

PROBE PLOT FOR 0·12 mm DIA Rh PROBE IN AN Ar PLASMA AT 300 mtorr 2 AMPS RMS.

Fig. 7.8 is a plot for a 0.12 mm diameter single rhodium probe placed in an argon plasma at 300 mtorr pressure 2 Amps rms. This shows that an increased current is drawn with a larger probe. Also, the floating potential value has changed. The I/V result shown in Fig. 7.7 under the same conditions but using a 50 μm diameter Pt/Rh single probe shows a positive floating potential to ground V_f of +5V, whereas the I/V curve in Fig. 7.8 using a 0.12 mm diameter single Rh probe shows a floating potential of -13V.

The floating potential is assumed by an immersed body in the plasma and is the voltage drop across an insulator to a grounded substrate. The body floats to a potential so that the ion and electron charge flow is equal. The results show that the degree and sign of potential depends on probe size.

Single probes for use in a carbon tetrachloride plasma presented a particular problem because of the need to use a probe material which did not chemically react during the experimental runs. The 50 μm diameter Pt/Rh probes used in the CCl_4-plasma did not give consistent results and this may have been due to chemical etching of the platinum in the probe metal.

When a 0.12 mm dia rhodium probe was placed in a CCl_4 discharge an I/V characteristic was obtained as shown in Fig. 7.9. This differs from the results for a similar probe (0.12 mm Rh) obtained for an argon discharge under the same pressure/power conditions (0.3 torr pressure 2 Amps rms); a much larger current was drawn by the probe in the CCl_4 plasma. Also only a small positive bias could be applied to the probe ∿ (+2 volts) before it melted. Fig. 7.9 also shows the I/V plots at 0.3 and 1.0 Amps r.f. supplied to the plasma from the generator.

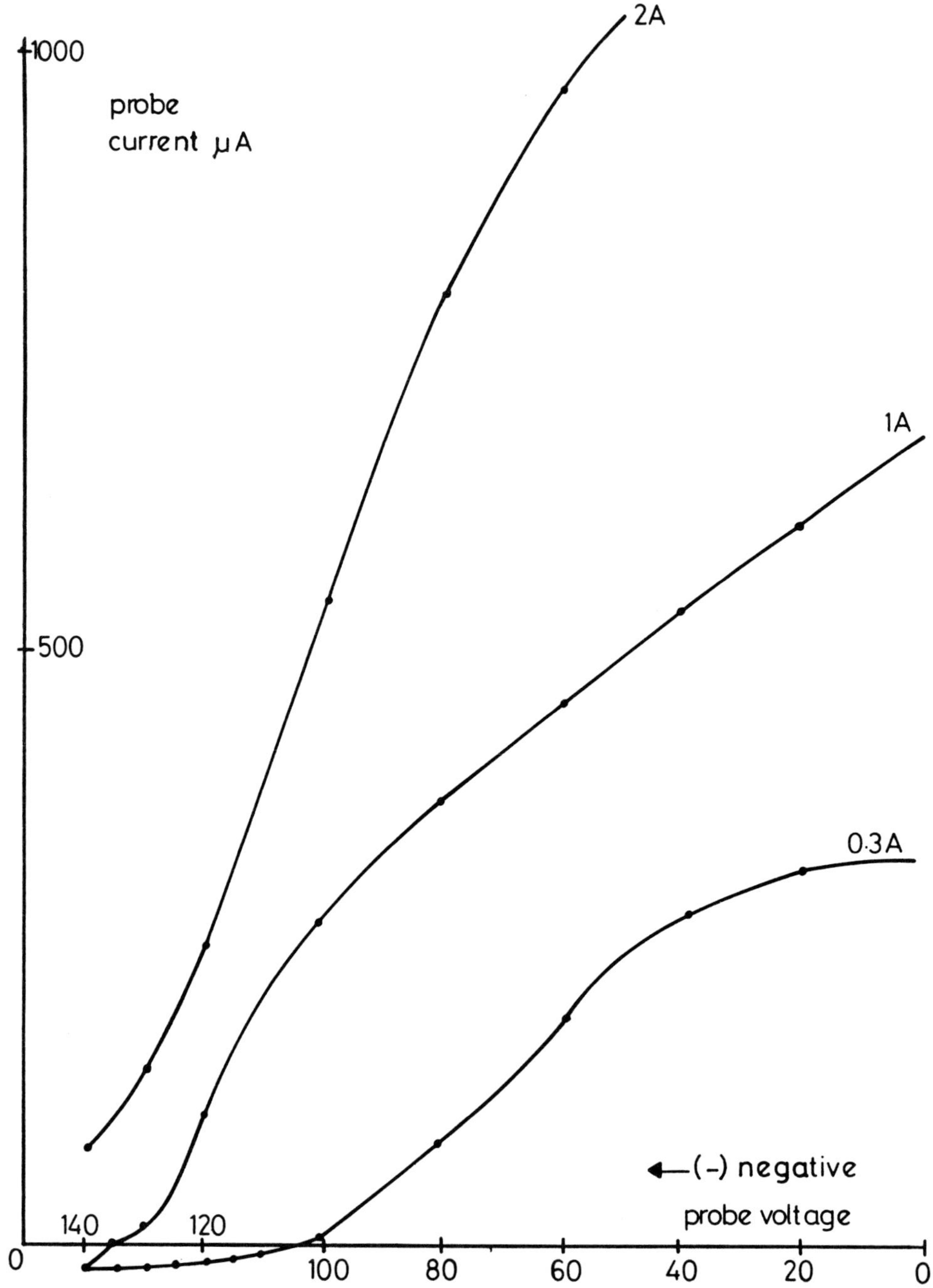

FIG 7.9

PROBE PLOT FOR 0·12mm DIA Rh PROBE IN A CCl_4 - PLASMA AT 300 mtorr PRESSURE AND VARIOUS CURRENT LEVELS

7.3.3 Symmetrical double probe study

The reference electrode in the single probe work was the grounded vessel and thus the system can be classed as a highly asymmetrical double probe method. Double probes can be considered as a variation of the single probe method where the reference electrode instead of being the ground surface of the chamber in contact with the plasma, is another probe and in a special case of equal area.

The double probes used in this study were in one case made of Pt/Rh 50 μm dia. by 4 mm length and in the other the double probe was Rh 0.12 mm dia by 4 mm length. The probe pair were 2 cm apart and situated midway between the electrodes. Each double probe was fabricated, pre-cleaned and installed in the same way as the single probe. The measuring circuitry is shown in Fig. 7.10 with 1 mH chokes connected to each probe to provide high impedance to the r.f. excitation component. Co-axial cable was used throughout. The probes were biased with respect to each other but electrically isolated so that they floated in the plasma.

Fig. 7.11 shows double probe I/V plots for an argon plasma at 0.1, 0.3 and 0.5 torr pressure at an r.f. plasma supply current of 2 Amps rms. The probe currents were much smaller than those obtained in the single probe case for equivalent probe and plasma conditions. Since the probes were identical and floating the total external current drawn from the plasma by the negatively biased probe will rise to saturation and equal the total positive ion current drawn by the probe external sheath area. The positive ion current is small compared to that which could flow as electron current if it were not limited by the counter flow of the positive ions.

Thus with double symmetrical probes one should obtain a symmetrical V/I characteristic with saturation currents at positive or negative biases being equal and positive ion current limited. The double probes used in the experiments discussed here did not suffer from r.f. interference to the same extent as the single probes. This is because when a single probe is maintained at a constant bias in a low frequency r.f. plasma between 1-900 kHz the plasma potential will vary with time.

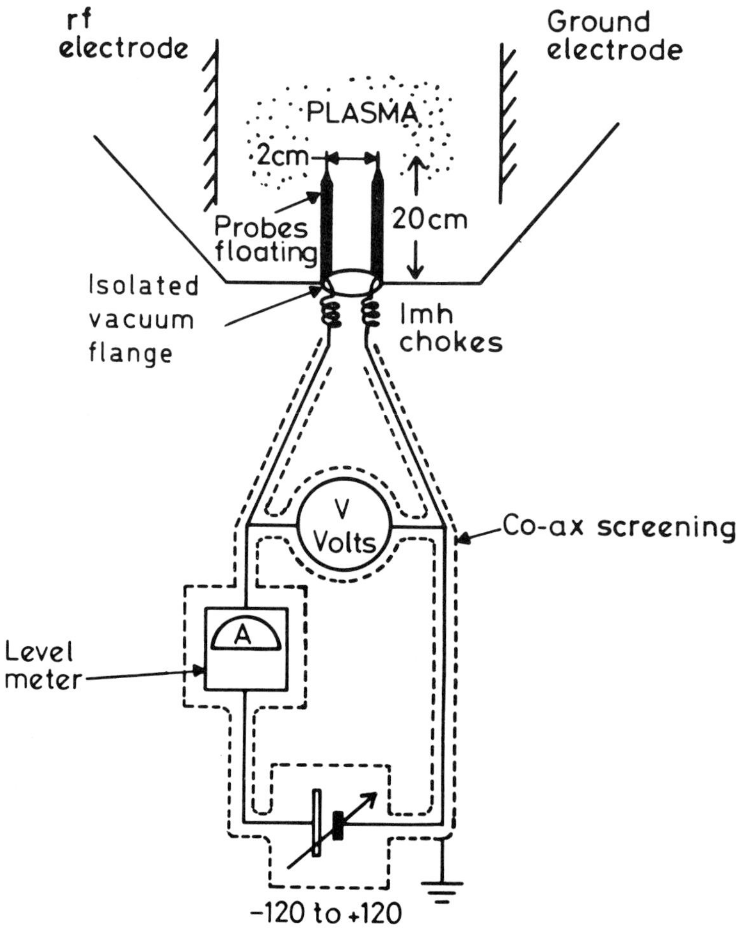

FIG 7.10

SYMMETRICAL DOUBLE PROBE EXPERIMENTAL ARRANGEMENT FOR PLASMA DIAGNOSTICS

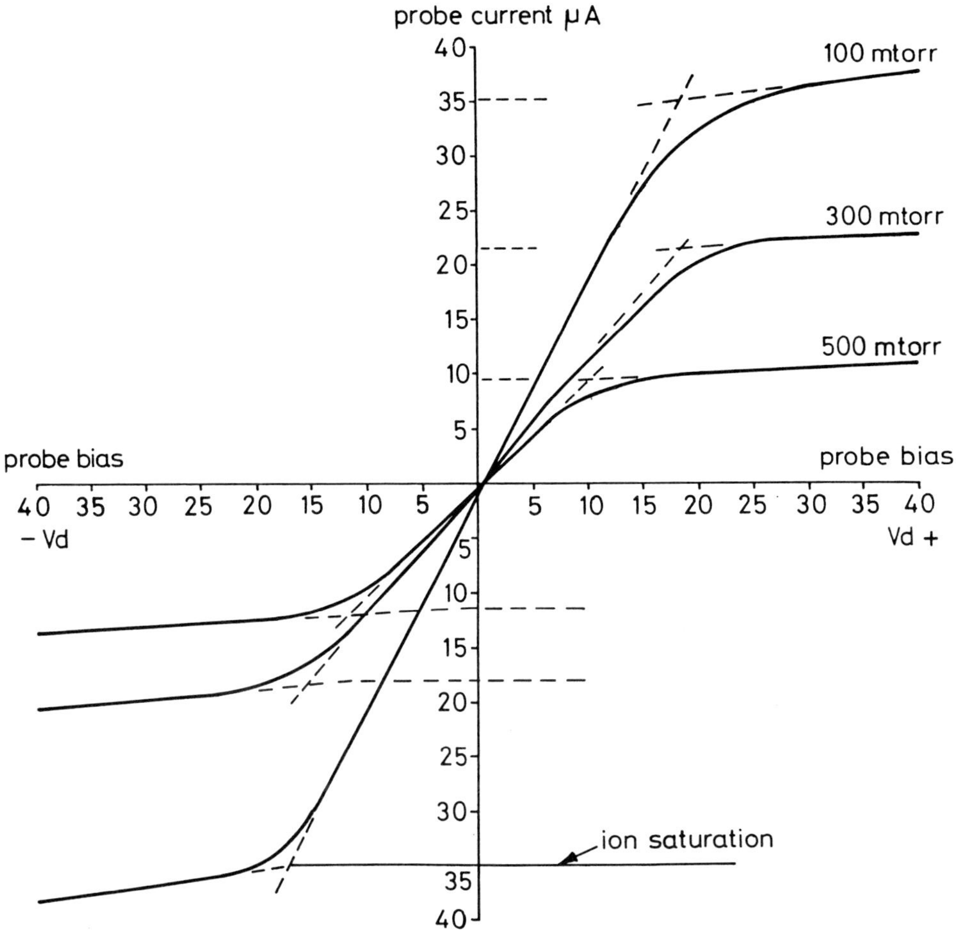

FIG 7.11

DOUBLE PROBE PLOT FOR AN Ar PLASMA USING 50μm DIA Pt/Rh PROBE AT PRESSURES 100, 300 AND 500 mtorr.

Thus the probe potential varies with time thereby giving a perturbation. With closely spaced double probes, however, both probes float in the plasma and the perturbation difference between them is almost zero.

Fig. 7.12 shows a double probe I/V plot of an Ar plasma at 0.3 torr pressure and at external supply currents of 3 Amps and 1 Amp rms

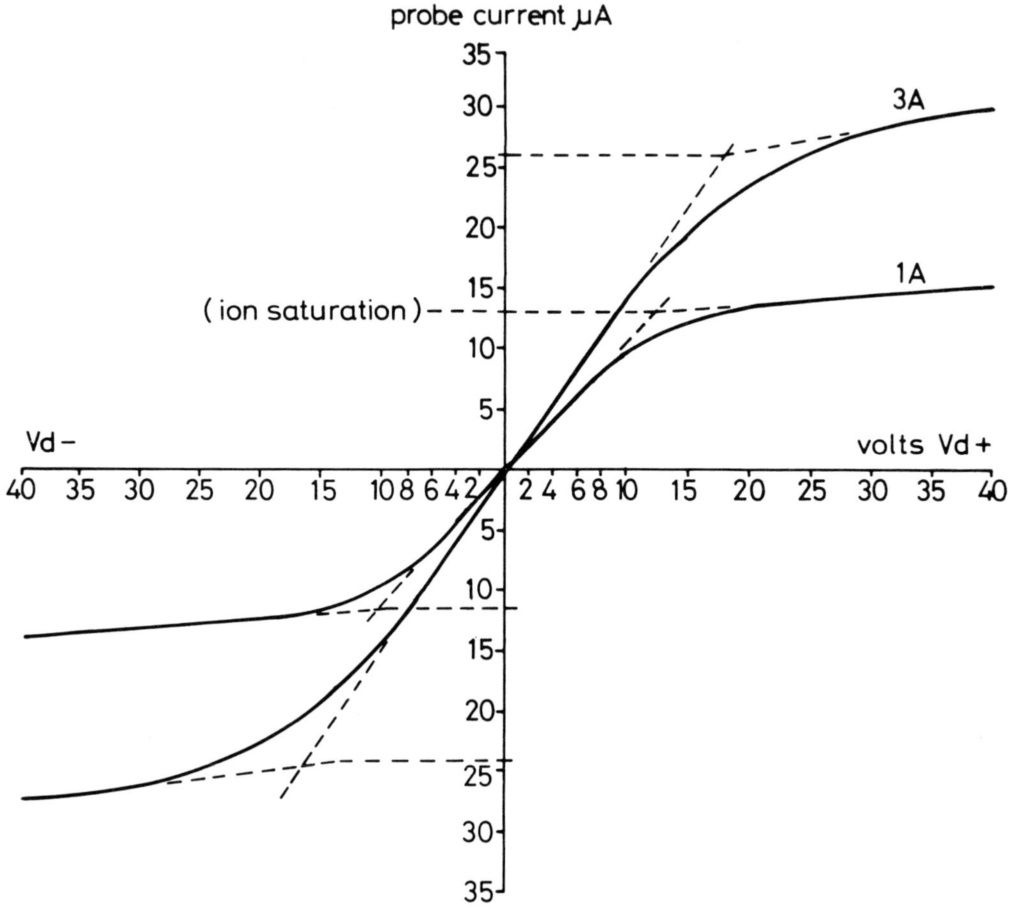

FIG 7.12

DOUBLE PROBE PLOT OF AN Ar PLASMA AT 300 mtorr PRESSURE AT 3 AND 1 AMP rms USING 50 μm DIA Pt/Rh PROBES

for the 50 μm diameter Pt/Rh probes. Fig. 7.13 shows a double probe I/V plot of an Ar plasma at 2 Amps rms, 0.3 and 0.5 torr pressure for the 0.12 mm diameter 4 mm long Rh probes.

Fig. 7.14 is a I/V double probe plot for a CCl_4 plasma at 0.3 torr and 2 Amps rms when using 0.12 mm diameter rhodium probes 4 mm long.

The probe currents are larger for the CCl_4 plasma than for the Ar plasma (Fig. 7.13) for the same diameter probes and the same conditions of pressure and current.

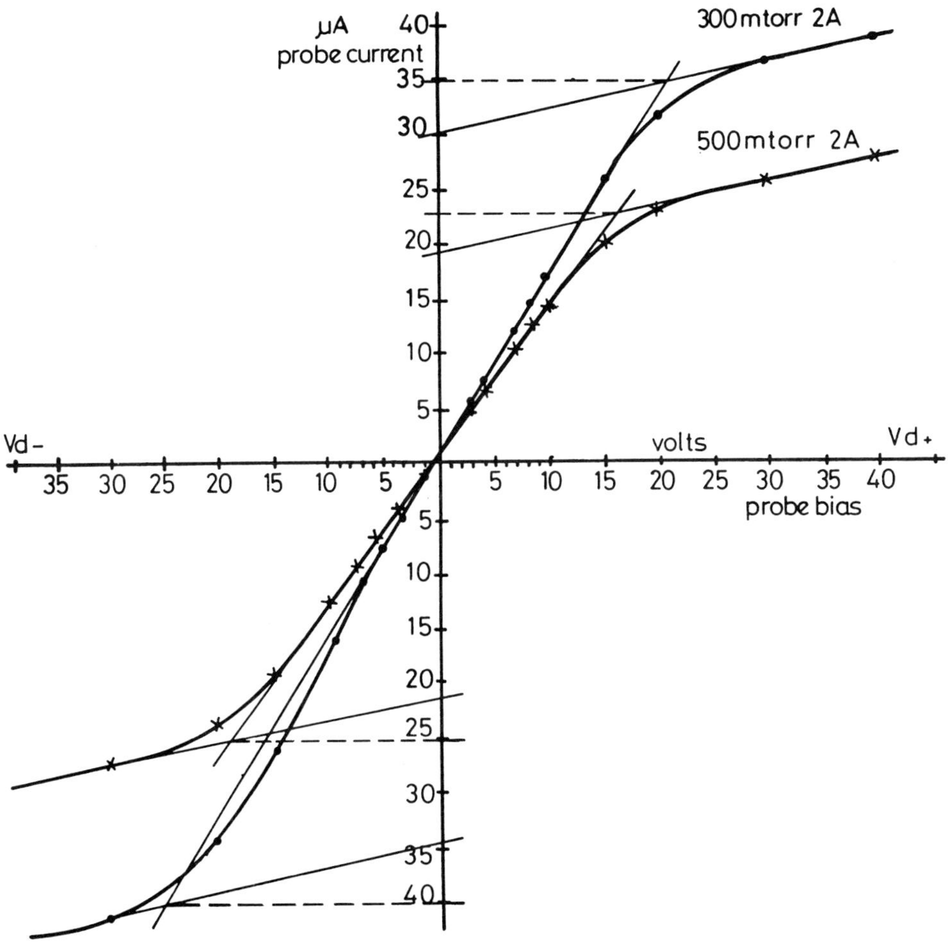

FIG 7.13

DOUBLE PROBE PLOT OF AN Ar - PLASMA AT 2AMPS RMS 300 AND 500 mtorr PRESSURE USING 0·12mm DIA Rh PROBES.

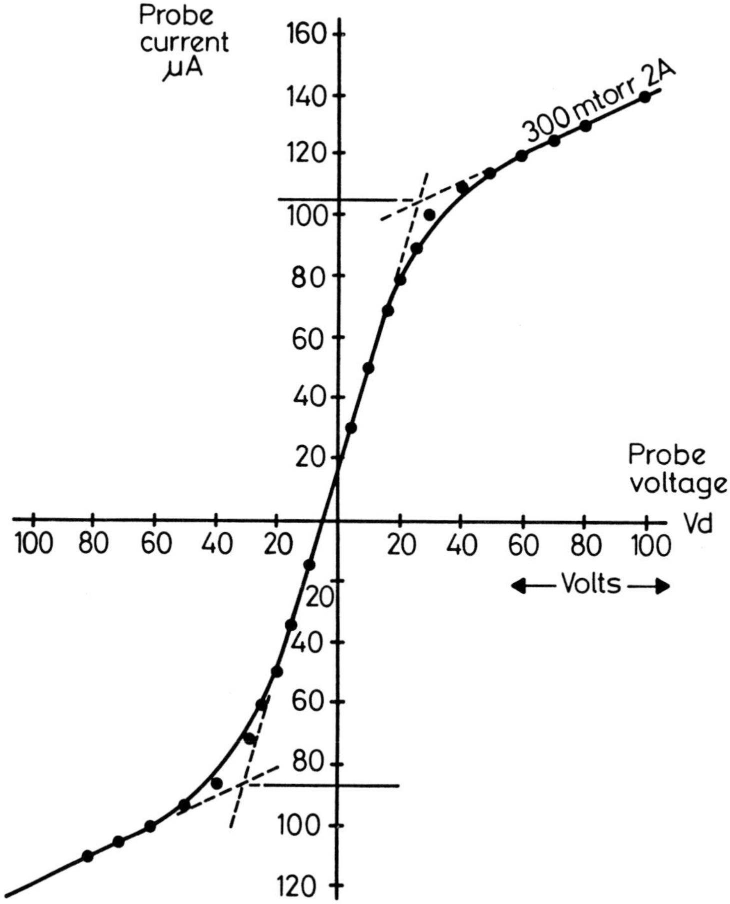

FIG 7.14

DOUBLE PROBE PLOT FOR A CCl_4 PLASMA AT 300 mtorr AND 2 AMPS RMS USING 0.12 mm DIA Rh PROBES.

Fig. 7.15 gives the double probe plot for a CCl_4 plasma at a current of 2 Amps rms and at pressures of 0.1 torr and 0.5 torr, using the same probes. There is now some sign of asymmetry in the double probe characteristic for the CCl_4 plasmas, i.e. values i_{es} for each probe are not equal. Also the change to the linear ion saturation regions of

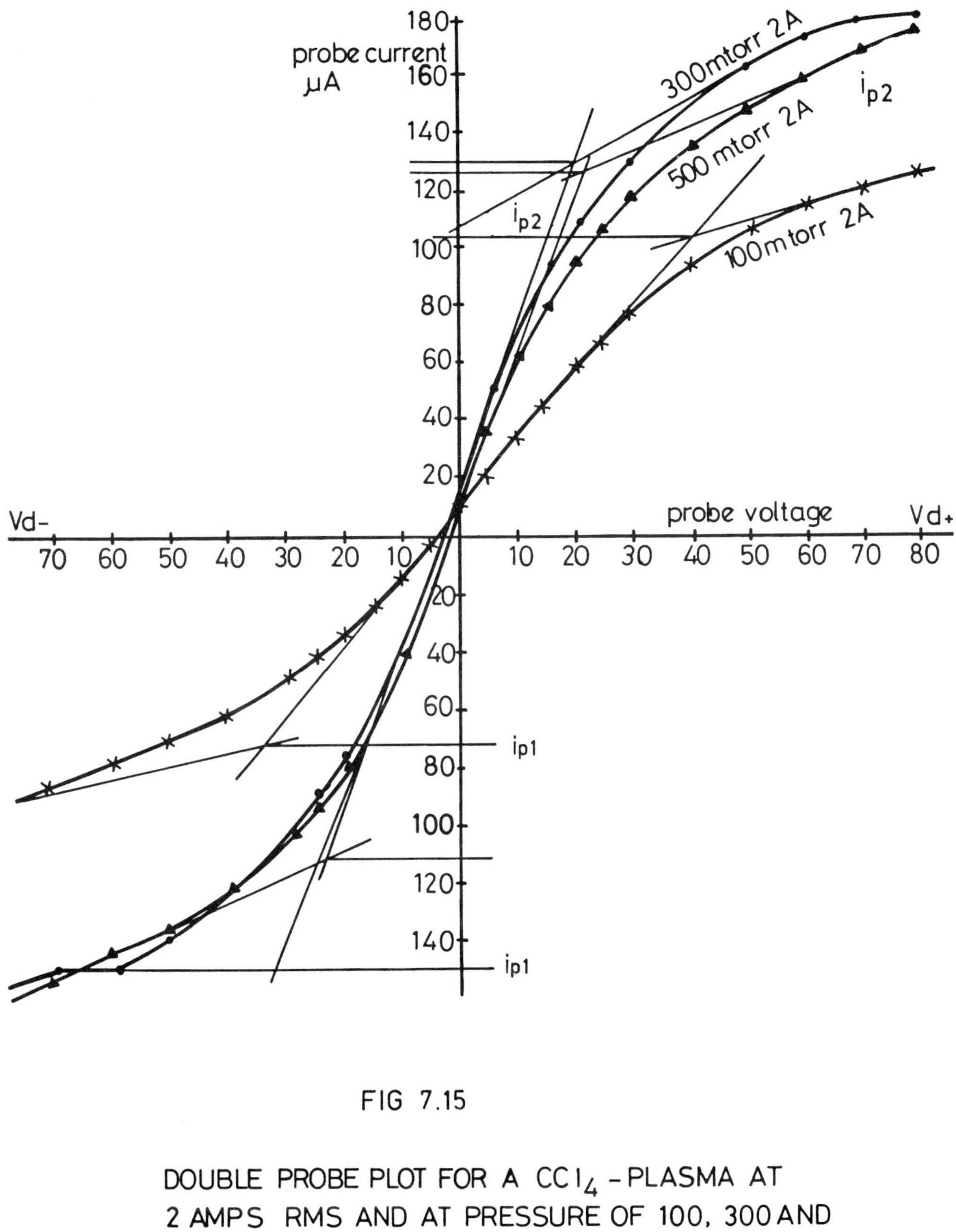

FIG 7.15

DOUBLE PROBE PLOT FOR A CCl_4 - PLASMA AT 2 AMPS RMS AND AT PRESSURE OF 100, 300 AND 500 mtorr, USING 0·12mm Rh PROBES.

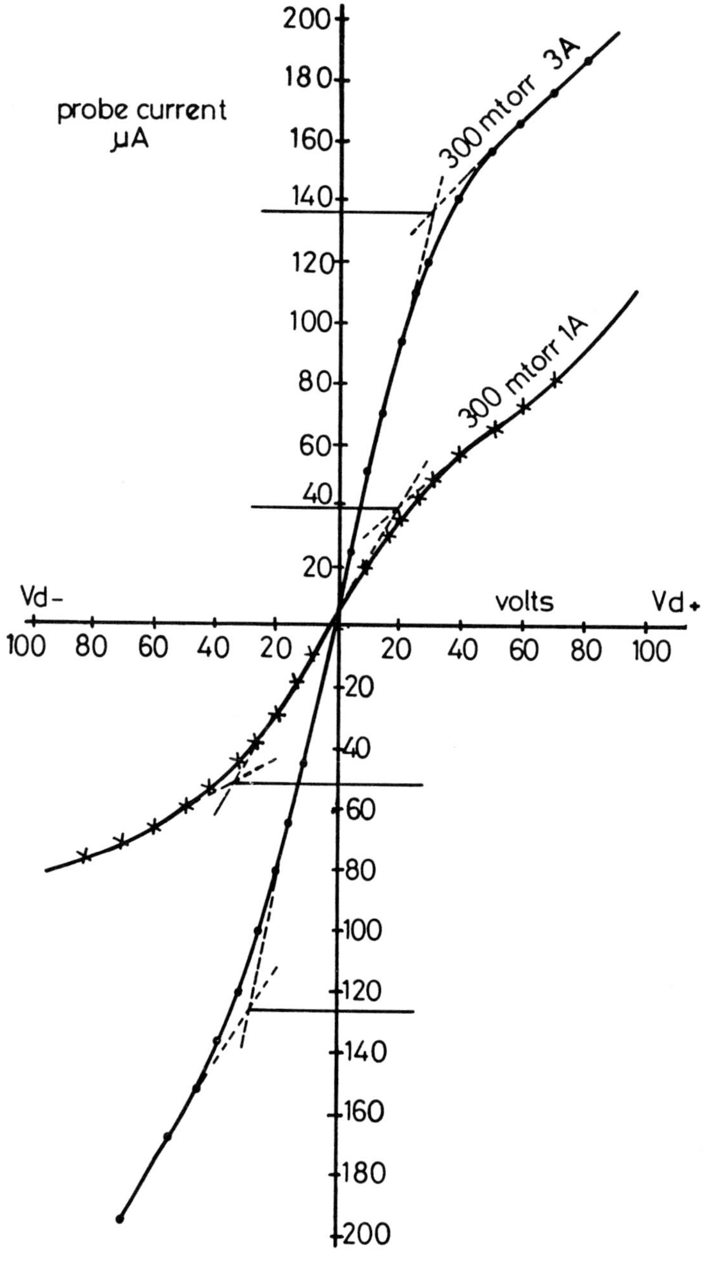

FIG 7.16

DOUBLE PROBE PLOT FOR A CCl_4 PLASMA AT 300mtorr PRESSURE AT 1 AND 3 AMPS RMS USING 0·12 mm DIA Rh PROBES.

the characteristic is less defined in comparison with the Ar plasma probe characteristics. Fig. 7.16 is a double probe plot for CCl_4 plasma 0.3 torr pressure and 1 and 3 Amps rms using 0.12 mm dia 4 mm long Rh probes.

7.4 Results and Discussion

7.4.1 Single probe work

Analysis of the single probe results under the conditions used in this work to measure the plasma potential was not possible because of the large currents drawn by the probe. Rundle et al (139), investigating an O_2-dc plasma at 1 torr pressure with a 1 mm dia Pt probe, also obtained large probe currents. They did not observe a knee in the I/V characteristic which would normally occur at the plasma potential. Also, they observed that the result was independent of the probe size or shape.

Thus the single probe work, presented in this chapter, by the author can only yield electron temperature and ion saturation data. This is because the single probes in the experiments here became hot and then gave a linear characteristic at small positive values. Hence measurements were always taken from the almost linear ion saturation region (i_{is}) at negative bias to ground through the exponential current rise region up to about +10 volts bias on the probe.

When the probe is positively biased, the current collected by the probe has a positive ion saturation current component which is neutralized by a component of the electron flow, but this becomes negligible compared with the electron current (i_e) as the probe to plasma potential is reduced.

Shown in Fig. 7.7 is a single probe plot of i_e (probe current) against probe voltage (V_B) for an Ar-plasma at 100 mtorr using 2 Amps r.m.s. supply current. The probe used at these conditions was a 50 μm diameter Pt/Rh-wire 4 mm long. At these conditions using a small diameter probe r_p at a pressure of 100 mtorr means that $r_p \leq \lambda_D$ where λ_D is the Debye screening length. Under these cirumstances the probe current is limited by orbital motion (O.M.L.)(134) and the relationship between the probe current and the probe voltage is:

$$i_e^2 = (\frac{1}{4} N_e e\bar{v} Af)^2 \tag{15}$$

where A is the probe surface area and f is the O.M.L. factor and is represented as

$$f = \frac{2}{\sqrt{\pi}}\sqrt{\frac{e V_s}{kT_e}} + 1$$

The equation (15) is similar to equation (5) since $\frac{1}{4} N_e e\bar{v} = J_{es}$ and $J_{es}A = i_{es}$. Using the data from Fig. 7.7 and plotting i_e^2 versus probe voltage, a straight line was obtained, shown in Fig. 7.17. The linear relationship between i_e^2 and probe voltage (V_B) at these conditions is consistent with the orbital motion current collection over the voltage range. The slope (S) of the straight line is

$$S = \frac{2}{\pi^2} \frac{e}{M_e} (N_e e)^2 \tag{16}$$

where M_e is the electron mass. This is the same slope value as in equation (5) using the i_{es} value. The intercept of this straight line of slope S with the V_B-axis is found by putting $i_e = 0$ in equation (15). It yields a sheath potential V_s, more negative than the plasma potential with

FIG 7.17

AR-PLASMA 2 AMPS 100mtorr, 50μm Rh/Pt SINGLE PROBE (PROBE CURRENT COLLECTED)² AGAINST PROBE VOLTAGE.

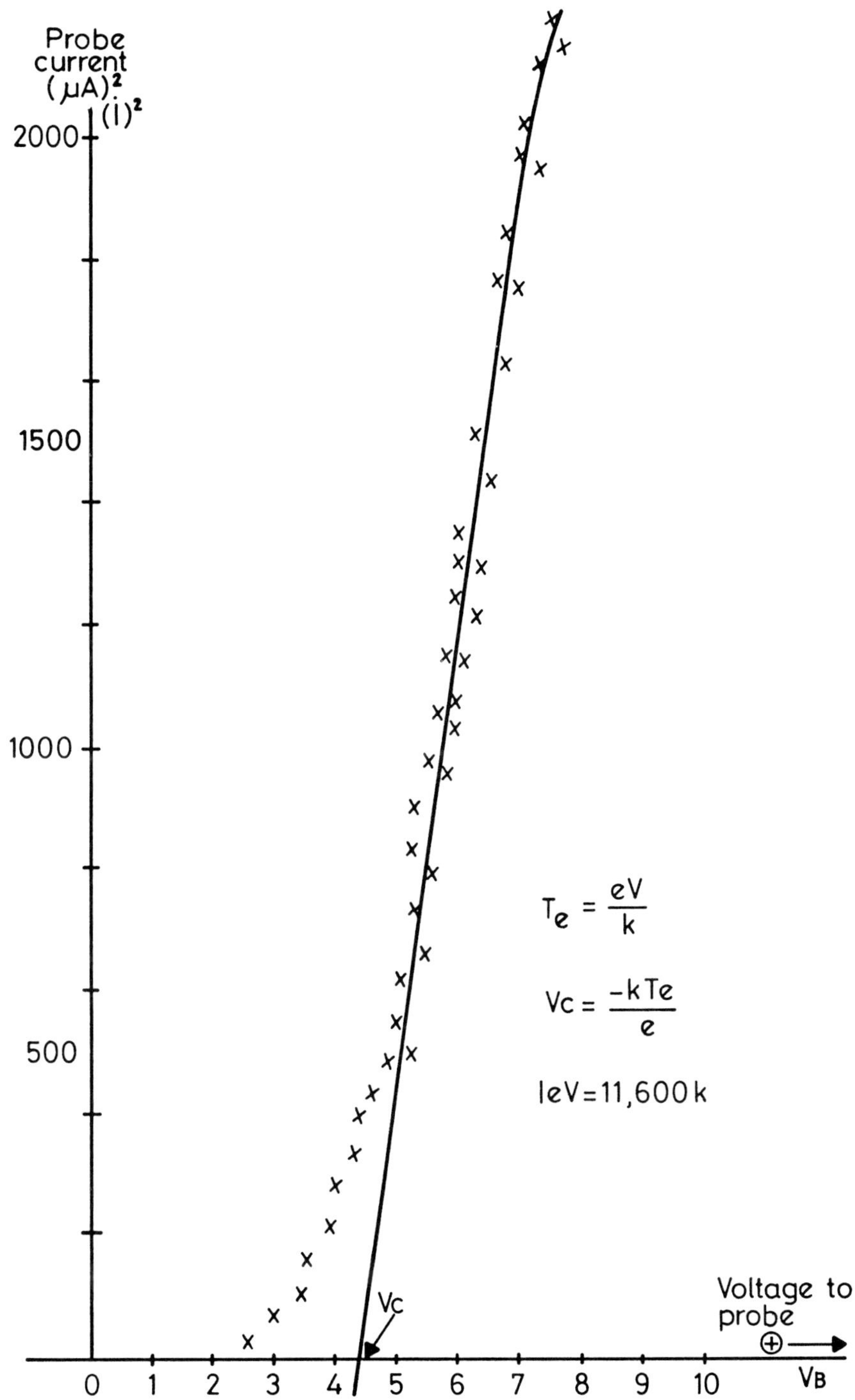
Probe
current
(μA)²
(i)²
2000
1500
1000
500
0
1
2
3
4
5
6
7
8
9
10
Vc
Voltage to
probe
⊕
VB
Te = eV/k
Vc = −kTe/e
leV=11,600k

respect to ground i.e. $V_B^{(i_e = 0)}$ a few volts positive probe bias and is equal to $-\frac{kT_e}{e}$.

$$V_B^{(i_e = 0)} = -\frac{kT_e}{e}$$

This is the same equation as (7) where

$$T_e = \frac{eV_c}{K} \qquad V_c = V_B^{(i_e = 0)}$$

This allows quantitative measurement of T_e without accurate knowledge of the space charge diameter surrounding the probe or the i_{es} value. The T_e value for Fig. 7.17 has been estimated to be 4.4 eV by this method. In principle T_e estimation could be made from using equation 5. However, i_{es} could not be determined here because the probe melted at below saturation current.

The initial curvature at low i_e^2 values shown in Fig. 7.17 could arise from the ion current to the probe. The ion collection at low i_e values can become a significant part of the probe current and may not contribute to the orbital-limited theory at these conditions.

Similar plots of $(i_e)^2$ versus V_B have been obtained by Woolsey et al. (140) with a dc-iodine plasma (weakly ionized) at 0.24 torr pressure. Also by Goodall and Smith (141) with He, Ar plasmas with diagnostics,done under afterglow conditions.

In the work here using 50 μm Pt/Rh single probes in an Ar-plasma, plots of i_e versus probe bias V_B for pressures above 100 mtorr did not give a straight line,probably because above this pressure the conditions are no longer orbital motion limited. The theory of Laframboise (137) predicts that for sufficiently small values of r_p/λ_D, i.e. $r_p < 3\lambda_D$ (where r_p = probe radius and λ_D is the Debye screening length) the electron or ion current collection is independent of the ratio of r_p/λ_D

and the current to the probe is orbital motion limited, assuming a Maxwellian distribution. In the work discussed here the probes used were 50 μm in diameter and λ_D can be estimated from the equation

$$\lambda_D = \left(\frac{2kT_e E_o}{N_e \; e^2}\right)^{\frac{1}{2}}$$

where E_o is the permittivity of free space, and N_e is the electron density. Hence, by estimating a value N_e to be about $5 \times 10^{17}\,m^{-3}$, λ_D would be 20 μm for the conditions used in Fig. 7.17 and thus current collection at the condition used borders on the O.M.L. regime*

The advantages of working in the OML regime is that one can have a small probe, which drawing small probe currents reduces plasma/probe perturbation. However probes in the OML regime achieve their best results usually under conditions of low pressure and low ion density. Some workers (142) still observe discrepancies between experiment and theory even when working in an Ar low density plasma afterglow condition.

From the I/V probe plots of Figs. 7.7 and 7.8 the semi log plots of ln i_e (probe current) against probe bias V_B (volts) were obtained and these are shown in Figs. 7.19 and 7.20. Each of the semi log plots reveal two regions. The T_e values estimated from the slope of the semi log plots were 6 eV, 12.4 eV for Fig. 7.19 and 8.9 eV, 12.8 eV for Fig. 7.20. The points where the slopes changed were around 8 volts for Fig. 7.19 and 10 volts for Fig. 7.20.

If these were real T_e values from two electron distributions in the plasma, then one would expect the lowest slope and thus the lowest T_e value to occur at the higher positive probe bias values, i.e. 12.4 eV and then 6 eV, because at a high positive probe bias the potential

*O.M.L. stands for Orbital Motion Limited.

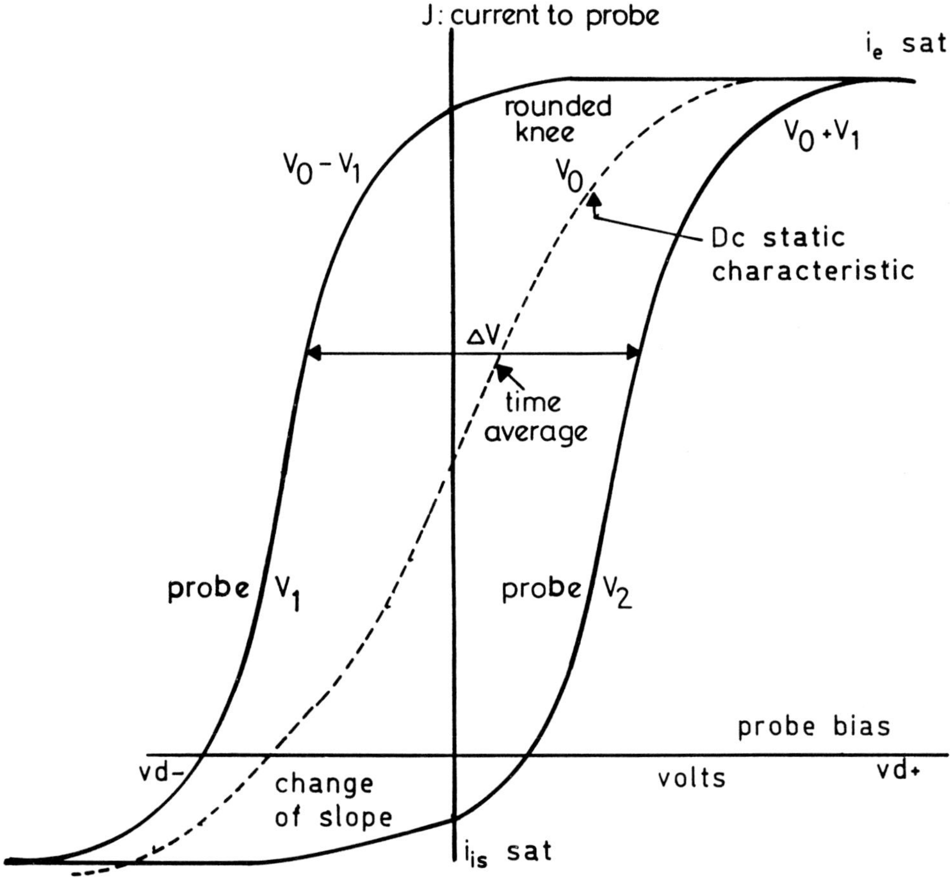

FIG 7.18

SCHEMATIC OF SINGLE PROBE I-V CHARACTERISTIC OF DC TIME AVERAGE CONDITION BY THE VARYING PROBE / IV PLOT. PROBE $I/V_1 \cdot V_1 = V_0 + V_1$ AND PROBE $V/_2$ IS $V_0 - V_1$ PRODUCING DC CHARACTERISTIC. ΔV VOLTAGE CHANGE FROM AC CYCLE CHANGES.
NOTE ROUNDED KNEE rf DISTORTION.

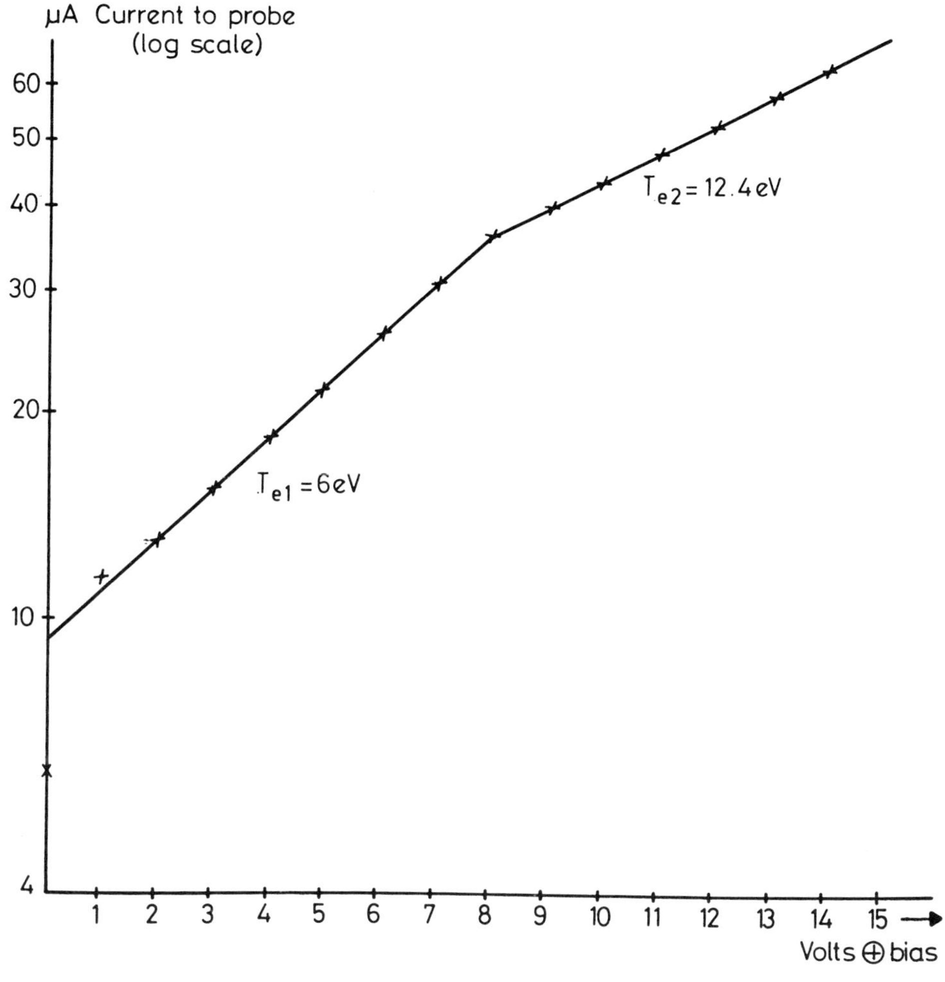

FIG 7.19

SEMI LOG PLOT OF A SINGLE PROBE CURRENT AGAINST POSITIVE PROBE BIAS (VOLTS) AR-PLASMA 300mtorr 2 AMPS USING 50µm Pt/Rh SINGLE PROBE.

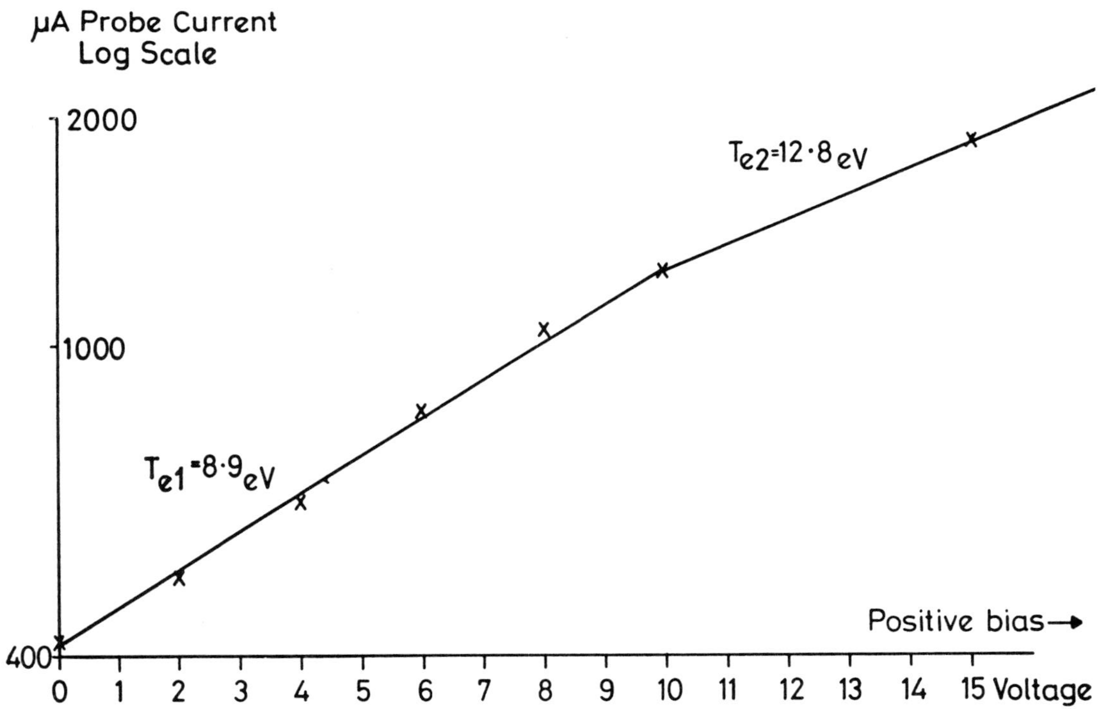

FIG 7.20

Ar PLASMA—300 m torr Rh 0·12 mm PROBE
SEMI LOG PLOT RESULTS TAKEN FROM
FIG. 7.8 LOG PROBE CURRENT AGAINST
PROBE BIAS (POSITIVE) FOR SINGLE PROBE

between the probe and the plasma is small and the lower energy electrons can then surmount the space charge barrier.

Using a probe at the work holder of an r.f. sputtering system, Holland and Priestland (171) found that the I/V curve showed two "knees" i.e. two plasma potentials and attributed this to the presence of electrons from the plasma and energetic electrons from an adjacent sheath. Thus, in systems which have two very different sheath potentials across the electrodes,such as r.f. and d.c. sputtering systems, it is possible to have two T_e values from two electron distributions. More recently, Esher et al. (143) using single probes in a sputtering system observed two electron populations of mean energy 1.6 eV and 0.43 eV in a d.c. discharge with +10 volts substrate bias at 80 mtorr Ar-discharge pressure.

In the r.f. planar system used here,the sheaths at both the top electrode and the workholder could have very high d.c. voltage drops possibly between ten to a hundred times higher than in some d.c. or r.f. sputtering systems. By operating a planar reactor at a pressure of 300 mtorr, the fast electrons leaving the sheaths can rapidly lose energy in the plasma. This may result in electrons with different distributions but of moderate energy.

However, there is another possible explanation for the slopes observed in Figs. 7.19 and 7.20. The curves could arise from the presence of the a.c./r.f. voltage waveform that exists across the plasma-probe junction. This causes the probe I/V characteristic to be distorted, so that it is an average of the I/V characteristic over many cycles. This distorted 'static' characteristic which is obtained in the presence of an r.f. voltage is a false characteristic. For example, assuming that the electron current collected is exponential with respect to the probe voltage, the instantaneous probe current i_{et} will be:

$$i_{et} = i_{es} \exp\left[-\left(\frac{e}{kT_e}\right)(V_{d.c.} + V_{a.c.} \sin \omega t)\right]$$

where $(V_{d.c.} + V_{a.c.} \sin \omega t) = \Delta V$ and is the sum of the d.c. and a.c. voltage components applied across the probe junction at time t.

The result of the averaging process is to impose a d.c. shift in the I/V characteristic, causing an apparent increase in the estimated T_e value and in the negative floating potential value to ground as observed in Figs. 7.7 and 7.8. The slope changes in the semi log plots i.e. Figs. 7.19 and 7.20 may also be associated with thermal emission effects as the probes began to glow for positive potentials to ground of 8-10 volts, thus increasing the current flow and changing the I/V slope. The effects of the a.c./r.f. averaging process on the probe characteristic are demonstrated in Fig. 7.18. The large (i_e) probe currents collected in a CCl_4-plasma at 300 mtorr pressure at various current levels are shown in Fig. 7.9. Fig. 7.9 shows the increase in the negative floating potential to ground, as the r.f. current to the discharge is increased.

Holland (172) using an r.f. planar reactor (13.5 MHz) has discussed the dependence on the position of the probe in a plasma. He considers that probes in the plasma volume can give negative floating potentials (V_f) with respect to ground, whereas one usually expects a positive value (V_f) for a plate probe adjacent to a ground electrode. The studies in his laboratory have shown that secondary emission of electrons from a probe exposed to electrons accelerated from an electrode sheath can cause local ionization and the probe becomes more negatively biased than the work plane, to limit electron flow.

The increase in negative floating potential shown in Fig. 7.9 with increasing applied r.f. current may be because of fast electron bombardment and by an increase in the a.c./r.f. distortion to the I/V characteristic.

Cantin and Gagne (144) have studied the effects of r.f. on probes and point out that current averaging is also responsible for an apparent increase in ion current collected. Hirsch (146), when using weakly ionized nitrogen in r.f. plasmas, also observed probe/r.f. interactions and errors in T_e by a factor of two, but no effect on the observed electron energy distribution. To minimise errors in the interpretation of single probe data obtained in r.f. plasmas, one should know or be aware of the a.c./r.f. component in the discharge.

Gill (132) has estimated the plasma potential rate of fall towards the ground potential in both high and low frequency plasmas. He has shown that in a high frequency plasma at 13.5 MHz excitation,the plasma potential is unable to fall as fast as the applied r.f. voltage. Thus the plasma potential skims along the peaks of the live r.f. electrode potential and causes a drop in the plasma potential to ground as well as a large d.c. negative bias to ground on the live electrode. It is possible with high frequency excitation to obtain good single probe characteristics with very little a.c./r.f. distortion as shown by other workers (136,152). This is because of the small plasma potential rate of change during a cycle at high frequencies.

In the experimental work here, using a planar low frequency reactor, single probes became overheated using static probe methods. An attempt was made to ramp a probe bias from -200 volts to +800 volts to ground in 0.5 second. The ramp voltage was used to obtain an I/V characteristic which it was hoped would be complete before the probe melted. The ramp generator was a simple high voltage cascade amplifier

FIG 7.21

AR-PLASMA 300 mtorr $\frac{1}{2}$ AMP RMS USING Rh PROBE 0.12 mm USING RAMP GENERATOR – 200 VOLTS RAMPED TO 600 VOLTS IN 2 SECONDS PROBE CURRENT mA vs PROBE BIAS.

incorporating high voltage low gain devices. However, when using the ramp generator the X-Y chart plotter limited the voltage ramp rate to 2 seconds due to the servo response and the current design limit of was exceeded. An attempt with the ramp generator using a low current discharge of 0.5 Amps rms in Ar at 300 mtorr is shown in Fig. 7.21, where no i_{es} was observed and the probe was damaged. The small peak at 150 volts is believed to be caused by the ramp generator electronics as it occurs at the same point irrespective of the experimental conditions.

A diagnostic method that can be used with an afterglow state is a rapid I/V sweep in a period of time shorter than the discharge decay time, i.e. with the r.f. plasma off, thereby avoiding interference. This method has been used by Goodhall and Smith (141) and by J. Waymouth (147) in low pressure and low ion density plasmas. However, these are not practical for high pressure, high charge density plasmas where recombination and decay are rapid.

In the experiments presented here, the use of single probes appears to be impossible for obtaining plasma potentials. Obtaining T_e values from single probe measurements requires mathematical manipulation of the data, some of which is subject to errors. For this reason most of the T_e data has been taken from measurements using the double probe method.

7.4.2 Double probe work

As discussed previously, single probes in low frequency plasmas are biased with respect to a reference electrode and are also subjected to an additional r.f. component. This can distort the I/V characteristic because the plasma potential is varying with time. Thus for a conventional single probe which is at a constant bias the probe potential is also varying with time. However, with symmetrical double probes located in the plasma, this effect is far less. This is probably because the entire probe system floats in the plasma, which is the reference potential. If the plasma potential V_p changes,then the floating potential V_f also changes in the same direction and thus the space charge sheath across the probes remains the same during the r.f. cycle, i.e. $V_s = V_p - V_f$.

Since double probes are floating with respect to the plasma,they will have smaller ion sheaths during measurements. This condition may help the collisionless probe criteria particularly for CCl_4 plasmas with short mean free paths for charged particles. Double probe I/V plots, using two 50 μm diameter Pt/Rh probes,have produced good symmetrical plots in an Ar plasma as shown in Fig. 7.11 at 2 Amps rms discharge current, at pressures of 100, 300 and 500 mtorr pressure. These plots show a decrease in ion saturation to the probes as the pressure increases. Using the same probes,Fig. 7.12 are plots for an Ar discharge at 300 mtorr pressure, at a discharge current of 1 and 3 Amps rms, showing the increasing ion saturation value as the current is increased.

The semi log plot of the probe data from Fig. 7.11, i.e. ln Γ where $\Gamma = \frac{2i_{is}}{i_{e2}} - 1$ plotted against the voltage applied between the probes (V_B) is shown in Fig. 7.22 yielding three straight lines from which T_e has been calculated to be 8.7 eV for 100 mtorr, 8.0 eV for 300 mtorr and 4.4 eV for 500 mtorr pressure. Thus a decrease in T_e with increasing pressure was observed.

Double probe plots in a CCl_4 plasma using the Rh probes 0.12 mm diameter 4 mm long are shown in Figs. 7.14 (at 300 mtorr pressure, 2 Amps rms current), 7.15 (2 Amps rms at 100 mtorr, 300 mtorr and 500 mtorr pressure) and 7.16 (300 mtorr at 1 and 3 Amps rms). The ion current saturation values are larger in CCl_4-plasmas than in Ar-plasmas for the same conditions of pressure, current and probe size. The plots obtained from CCl_4-plasmas appear to show some asymmetry in the characteristic making

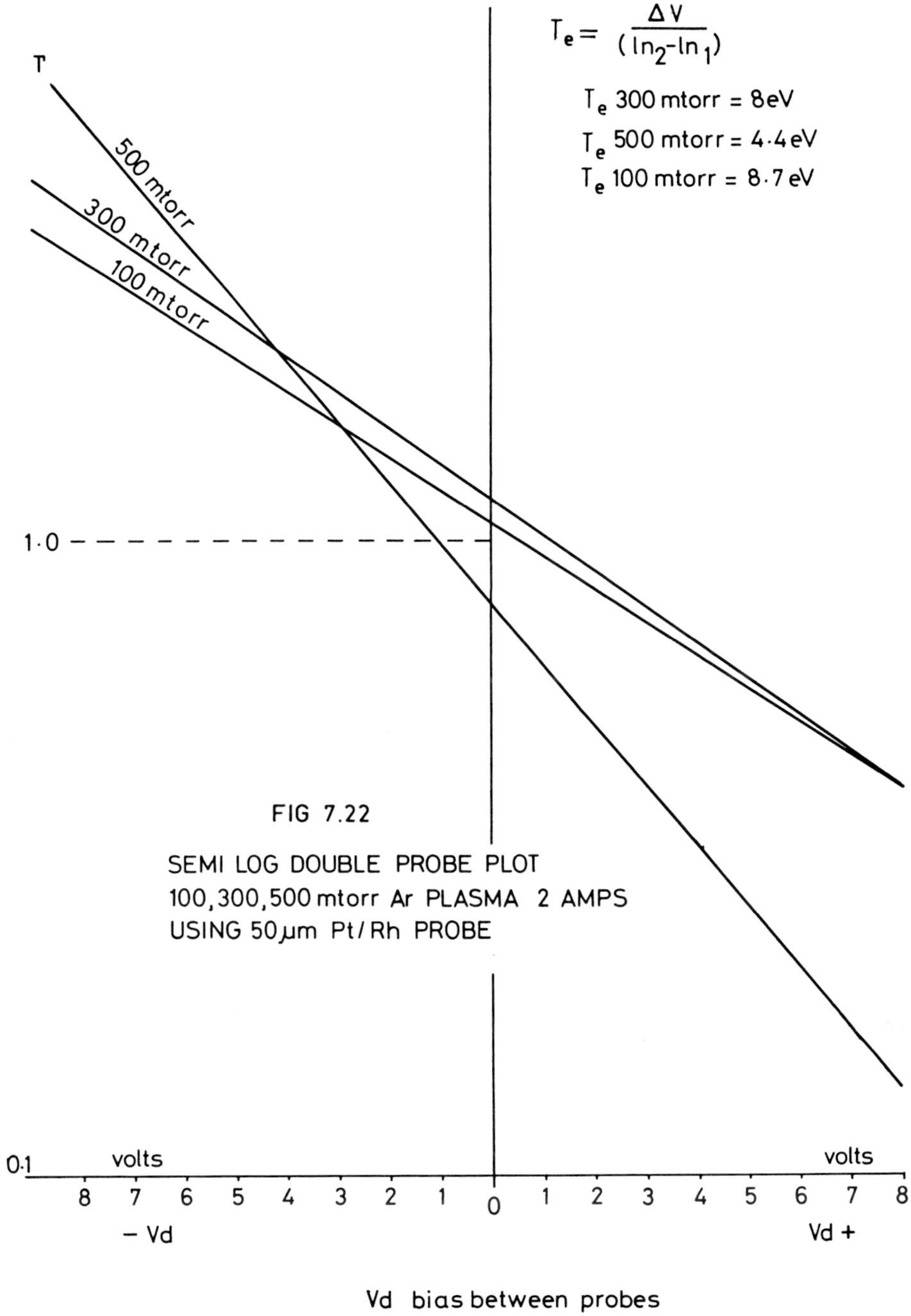

FIG 7.22

SEMI LOG DOUBLE PROBE PLOT
100,300,500 mtorr Ar PLASMA 2 AMPS
USING 50µm Pt/Rh PROBE

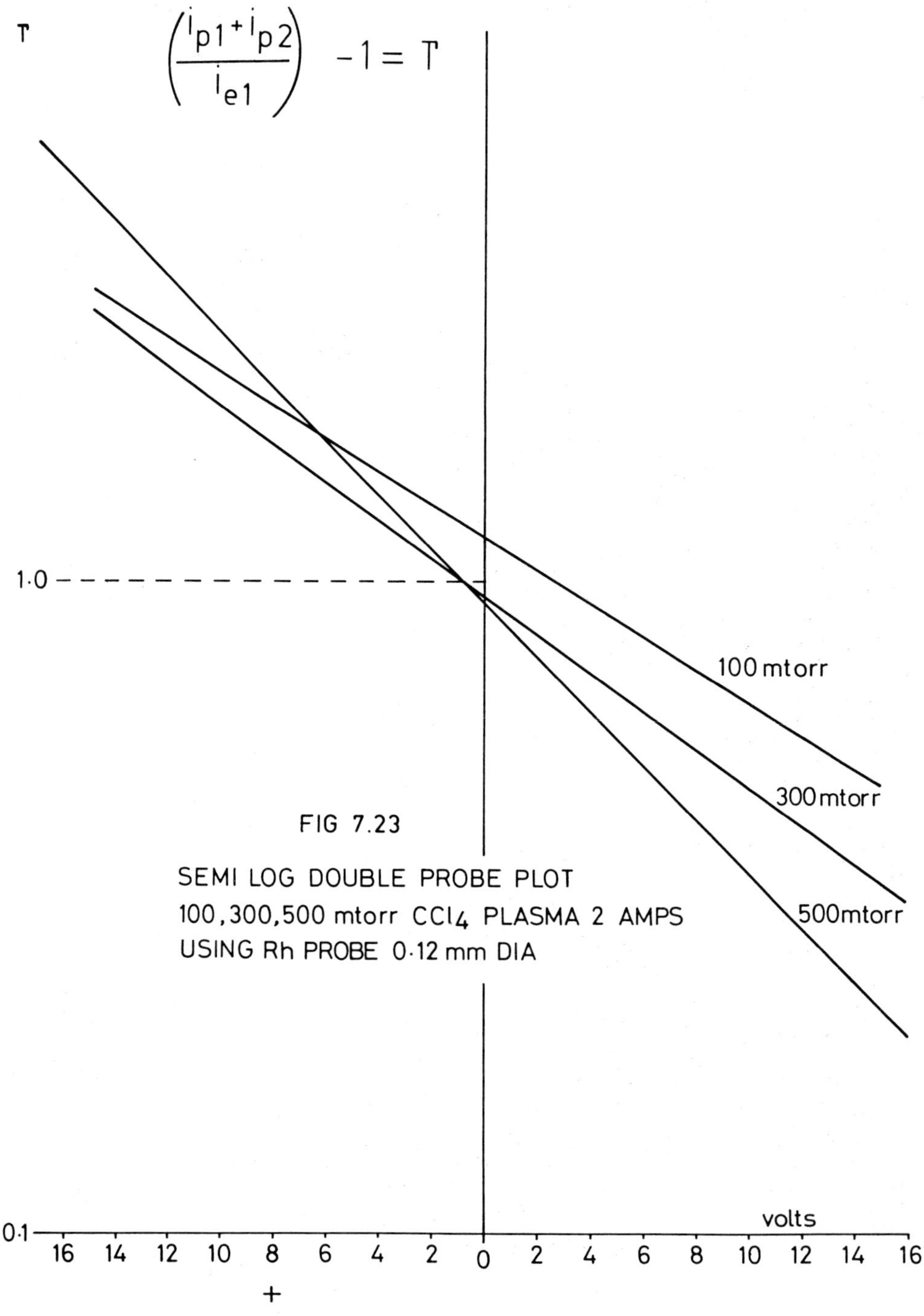

FIG 7.23

SEMI LOG DOUBLE PROBE PLOT
100,300,500 mtorr CCl_4 PLASMA 2 AMPS
USING Rh PROBE 0·12 mm DIA

it difficult to derive i_{is} values from the double probe I/V plots. This gives a larger deviation from the straight line semi log plots compared with the Ar-plasma semi log plots.

Fig. 7.23 is a semi log plot using the data from Fig. 7.15, from which T_e values measured for 100, 300 and 500 mtorr pressure were 16, 14.7 and 10.2 eV respectively. By using the intercept method, the T_e measured values were 17.8, 13.4 and 10.1 eV respectively. Fig. 7.24 is the semi log plot of the data from Fig. 7.14 and Fig. 7.16, from which T_e measurements at 300 mtorr pressure in a CCl_4 plasma at 1, 2 and 3 Amps rms current gave values of 10.0, 13.5 and 13.6 eV respectively. By using the intercept method, T_e values of 12.8, 13.4 and 14.3 eV respectively, were obtained.

The values of T_e obtained were larger than some reported values (see ref. 136). This may be because of the high r.f. power input and sheath potentials needed in plasma reactors to generate sufficient active gas species and high ion energies for rapid anisotropic etching.

Ion densities (Ni) can be derived from the ion saturation current (i_{is}) in terms of maximum current density $J_{is} = i_{is}/A$, where A is the current collecting area. The random current flow to a sheath boundary of unit area is given $J_{is} = \frac{1}{4} N_i e \bar{v}_+$ where $\bar{v}_+$ is the positive ion mean velocity, N_i the ion density, e the electron charge.

When a collecting electrode and sheath are parallel and semi infinite the electrode current density at saturation will be that at the sheath boundary. However, the collection area of a cylindrical probe is smaller than that of the space charge boundary surrounding the probe. Thus the sheath collection area A is difficult to estimate and involves the solution of Poisson's equation for cylindrical co-ordinates (28). As the potential across the sheath must also be known this was difficult to measure as the plasma potential was not known.

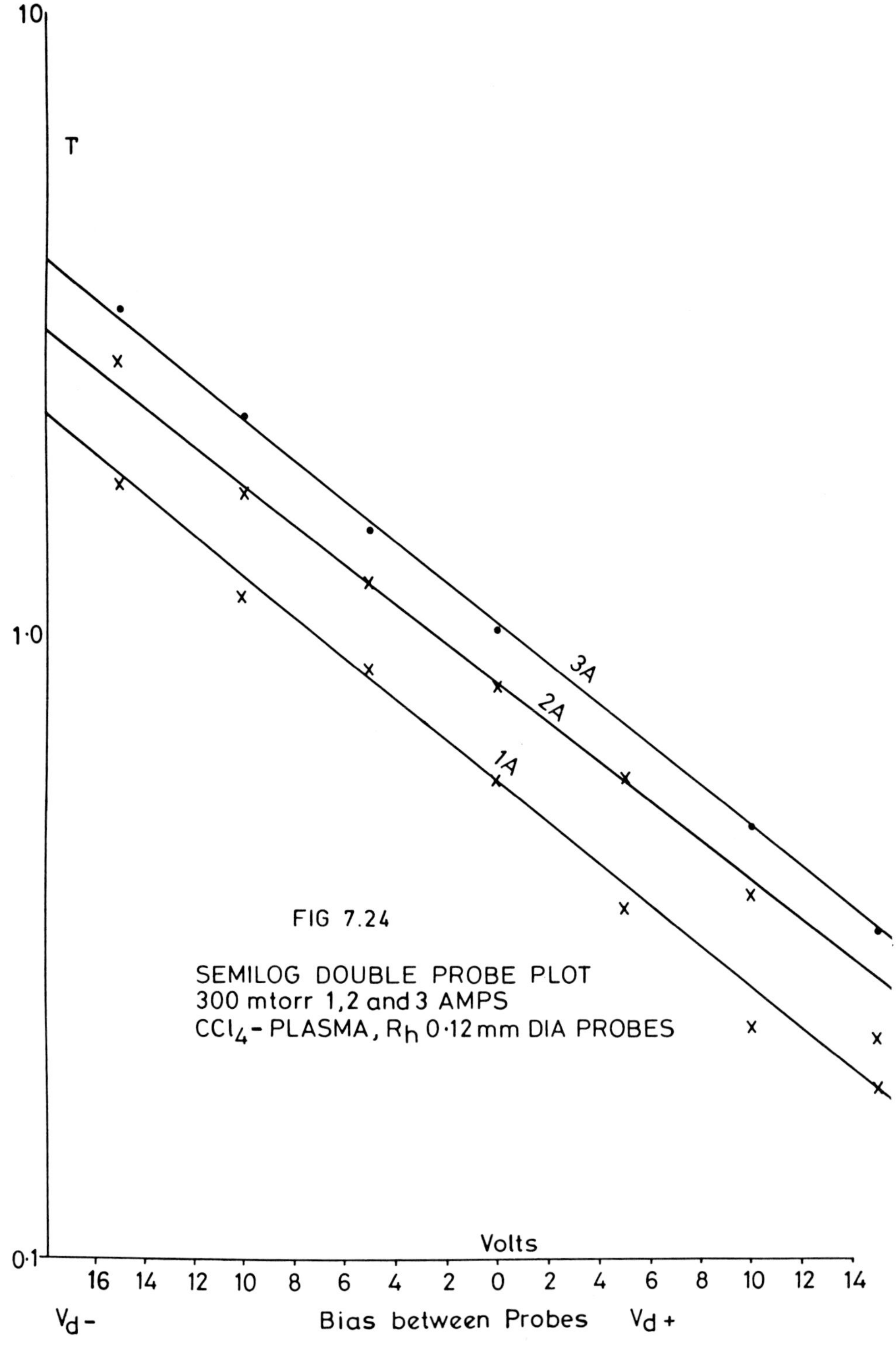

FIG 7.24

SEMILOG DOUBLE PROBE PLOT
300 mtorr 1,2 and 3 AMPS
CCl_4 - PLASMA, R_h 0·12 mm DIA PROBES

To overcome these problems, large 9 mm diameter disk probes were used and assuming the space charge sheath and probe collecting areas are almost equal then

$$N_i \simeq \frac{4 i_{is}}{A e \bar{v}_+}$$

where A = collection area of probe.

This equation also requires additional assumptions in estimating $\bar{v}_+$ from the ion kinetic energy. Thus the ion mass M_i was assumed to be that of singly charged atoms Ar^+ for an Ar-plasma and singly charged CCl_3^+ for a CCl_4-plasma. The ion temperature (T_i) was assumed to be approximately equal to the gas temperature T_g and thus $\bar{v}_+$ may be estimated from:

$$\bar{v}_+ = \sqrt{\frac{8}{\pi} \frac{kT_i}{M_i}}$$

The Ar-plasma will consist mainly of Ar^+ ions as the charge carrying species, however for a CCl_4 plasma the CCl_3^+ is not the only ion and may only be a large fraction of the ion current species.

Fig. 7.25 is a double probe plot for the 9 mm diameter aluminium disk probes at 2 Amps rms and 300 mtorr pressure in an Ar-plasma. Fig. 7.26 is the same plot with the Al-probes coated with a thin Rh film under the same conditions but in a CCl_4 plasma. Assuming a T_g value slightly above ambient of 400 K, the N_i value for an Ar-plasma was estimated to be $\sim 10^{17}$ ions m^{-3} and for a CCl_4-plasma to be $\sim 10^{18}$ ions m^{-3}. These values were obtained from Figs. 7.25 and 7.26 respectively. T_e values were also estimated by the intercept methods from Figs. 7.25 and 7.26 yielding 10 eV for the Ar-plasma and 16.7 eV for the CCl_4-plasma which is similar to the T_e values measured from other plots under the same conditions.

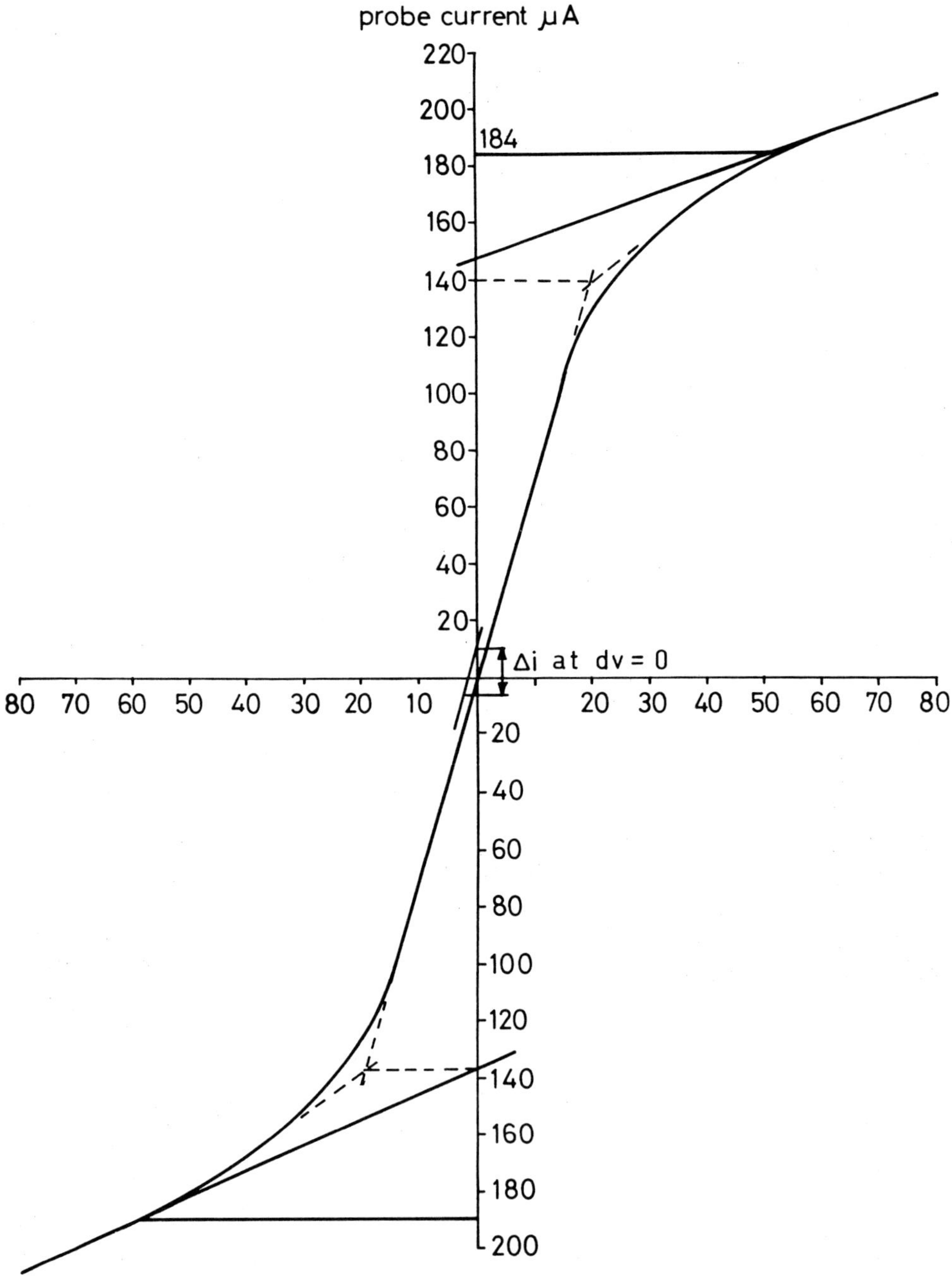

FIG 7.25

Ar PLASMA 2 AMPS 300 mtorr PRESSURE
USING LARGE 9mm DIA PROBES (Al) FOR
Ni DETERMINATION.

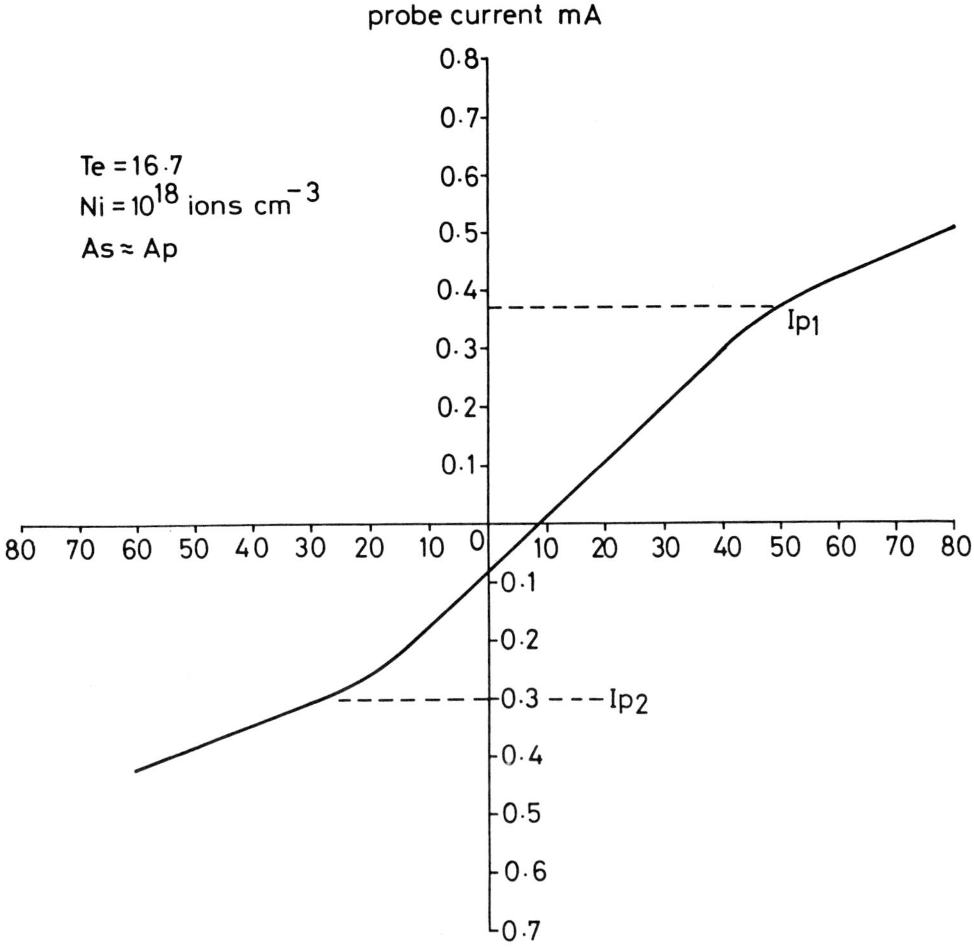

FIG 7,26

CCl_4-PLASMA 2 AMPS 300 mtorr PRESSURE USING Rh COATED Al PROBES 9mm DIA (Ni DETERMINATION)

Table 7.1 is a summary of the results obtained from the 50 μm diameter Pt/Rh probes in an Ar-plasma. All three methods for calculating T_e from the probe plots are in reasonable agreement. The electron temperature and the ion current saturation decreased as the pressure was increased in the Ar plasma with a constant current supplied to the plasma of 2 Amps rms. This could be because as the pressure increases the energetic secondary electrons leaving the space charge collide and lose energy near to the sheath. This may result in a localized high concentration of charge density just above the sheath boundary. Such an

Table 7.1

SUMMARY OF DOUBLE PROBE RESULTS Te MEASUREMENTS USING 50 μm DIA P/tRh PROBE 4mm LONG IN AN Ar-PLASMA

Plasma Conditions Pressure and Current	Te Semi Log Method	Te R Method	Te Intercept Method	Measured Ion Saturation Current to Probe μA
2 amps rms 100 mtorr	8.7 eV	8.7 eV	8.4 eV	35
2 amps rms 300 mtorr	8.0 eV	8.2 eV	7.2 eV	19.7
2 amps rms 500 mtorr	4.4 eV	5.2 eV	4.2 eV	10.5
1 amp rms 300 mtorr	5.2 eV	5.8 eV	5.0 eV	12
3 amp rms 300 mtorr	8.7 eV	8.7 eV	8.4 eV	25

effect could coincide with the observed intense glow just above the space charge sheath as the pressure is increased from 300 to 500 mtorr (67). With recombination events increasing at higher pressures it is possible that between the electrodes and around the vicinity of the probes the ion density will be lower. This will also lower T_e since the fast electrons entering the plasma would have lost more energy through more frequent collisions.

The results for the CCl_4 plasma using 0.12 mm diameter Rh probes are summarised in Table 7.2. They show the same trend as the Ar plasma of decreasing T_e with increasing pressure and decreasing supply current. However, the ion current saturation shows a very small rise and fall with decreasing pressure. There is an increase in ion current saturation as the supply current increases to the discharge. However accurate measurement of ion current saturation is difficult as there is no sharp change in slope in the plots.

There are several papers on electron energy changes with pressure.

Some examples are as follows:

E. Mainwaring and J. Swift (149) observed a decrease in T_e with an increase in pressure in a d.c. He-Ne discharge at 60 mtorr; A. Cantin and R. Grange (150) using an Ar-plasma, also observed an increase in T_e with a decrease in pressure; M. Kushner (151) observed localized T_e and N_i variations within the plasma volume using CF_4/H_2, CF_4/O_2 and Ar r.f. plasmas. For an Ar plasma at 80 mtorr Kushner observed a localized increase in T_e towards the electrode plasma boundary and the increase was larger if the ground electrode surface was bare metal and not an insulated surface. This is probably because the bare electrode surface was a poor secondary electron emitter thus larger potentials were required across the sheath to maintain the same charge density.

Table 7.2

SUMMARY OF DOUBLE PROBE RESULTS, Te MEASUREMENTS
USING 0.12 mm DIA Rh PROBE 4 mm LONG IN A CCl_4 - PLASMA

Plasma Conditions Pressure and Current	Te Semi-log Method	Te Intercept Method	Measured Ion Current Saturation Probe (Large Errors) μA
2 Amp rms 100 mtorr	16	17.8	90
2 Amp rms 300 mtorr	14.7	13.4	120
2 Amp rms 500 mtorr	10.2	10.1	110
1 Amp rms 300 mtorr	10.0	12.8	45
2 Amp rms 300 mtorr	13.5	13.4	100
3 Amp rms 300 mtorr	13.6	14.3	130

*Note R equivalent resistance method could not be used as the probe plots were asymmetrical and thus measurements would be difficult by this method. The asymmetry of the plots also make measurement of ion current saturation values difficult.

Kushner measured T_e between the live and ground electrodes in an Ar plasma at 80 mtorr, 280 volts peak-to-peak, to be 11 eV.

Probe position in a plasma is important as electrons flowing from sheaths can affect measurements (172,152) and probes can be used for mapping local changes near to space charge sheaths at targets in sputtering systems (153). Most workers place the probe midway between the excitation electrodes and away from the walls.

Thornton (136) has used double probes in a magnetron sputtering system at 1 mtorr Ar pressure and determined T_e to be 3.5 eV. The T_e values obtained for the planar reactor used in this study are higher than those in sputtering systems which have lower charge density and operating pressure.

The estimation of T_e from single probe work here was between 6-9 eV for an Ar-plasma 300 mtorr 2 Amps rms. Usually single probes sample the complete electron distribution and, from the I/V characteristic, it is evident whether or not the electrons are in a Maxwellian distribution. However, in the single probe measurements made here, the probe was only biased over a small voltage range, between -5 and +10 volts with respect to ground. This was because the probe became hot at about + 10 volts and therefore only the high energy tail of the electron energy distribution was sampled.

The T_e value of 8.0 eV was obtained from the double probe work here for an Ar-plasma 300 mtorr 2 Amps rms. Although the results using double probes were more reproducible than single probes, double probes are always negative with respect to the plasma and hence they will also sample only the high-energy tail of the electron energy distribution. If distribution is Maxwellian, then sampling small regions will give an average T_e for the system. However, the double probe method can easily ascribe a temperature to electrons even though they may be far from being thermalized.

During the work reported here, Vossen (14) published a paper giving the results of measuring the plasma potential (V_p) and floating potential (V_f) in a plasma planar reactor similar to the one used here. From Vossen's information and using Podgorngi's equation, discussed below, it should be possible to estimate T_e for Vossen's system and compare it with the results given here. As

$$J_e = \underbrace{N_e e \left(\frac{kT_e}{2\pi M_e}\right)^{\frac{1}{2}}}_{\substack{\text{random current} \\ \text{incident electrons}}} \quad \underbrace{e - \frac{e\Delta V}{kT}}_{\substack{\text{Boltzmann's} \\ \text{factor}}} \qquad (1)$$

where J_e is the electron current density, M_e is the electron mass and ΔV is the potential barrier for the electrons to surmount. At the floating potential, i.e. the potential at which the current density of electrons and ions are equal, the most energetic electrons must overcome a potential barrier of ΔV to provide an electron current to balance that of the ions. The ions incident at the plasma/sheath junction are all collected and reach the surface thus:

$$J_i = eN_i \frac{kT_i}{2\pi M_i} \qquad (2)$$

where J_i is the ion current density

T_i is the ion temperature

M_i is the ion mass (M_e is the electron mass)

N_i is the ion charge density which $N_i \approx N_e$ in plasmas

Since $J_e = J_i$ on a floating surface, equation (1) and (2) gives

$$\Delta V = \frac{kT_e}{2e} \ln\left(\frac{M_i}{M_e} \cdot \frac{T_e}{T_i}\right) \quad \text{(ref.28)} \qquad (3)$$

The electron temperatures have been estimated for both Ar and CCl_4 plasmas at various conditions. Using the Podgorngi equation one can determine ΔV, i.e. the voltage drop across the space charge sheath of a surface at floating potential. This will be considered further in the next section which reports on the r.f. voltage/current characteristics and related d.c. negative bias properties for the system studied.

7.5 Conclusions on Probe Uses

Single probes are thermally damaged when operated at a positive bias in a planar plasma etcher, because the degree of ionization in the reactor is high, resulting in a large electron flow at small positive potentials with respect to ground. Also single probes are unsuitable for T_e measurement because of an r.f. field time average effect occurring across the probe sheath. Double probes give reproducible T_e measurements for Ar-plasmas but to a lesser extent for CCl_4-plasmas. There is a decrease in T_e with increase in the operating pressure. The ion density has been crudely estimated under device fabrication conditions to be $\sim 10^{18}\, m^{-3}$, hence the large currents drawn by single probes.

These are very large T_e values probably because of the large negative V_f Vossen obtained from single probe work and it suggests the result is in some way in error. This could be because of the a.c. current averaging of the probe characteristic which he may not have taken into account. Vossen's probe managed to survive at large positive potentials to ground of 355 volts at V_p in an O_2-plasma probably due to very low current settings or by use of some type of rapid ramp generator. Because the single probes here could not survive positive potentials to ground of +10 volts, there may be a significant charge density difference between Vossen's system and that used here.

where $V = (V_p - V_f)$

which is the Podgorngi equation. Chen (145) uses an approximate expression for ion current density which does not involve T_i ;

$$J_i = N_e e \left(\frac{kT_e}{4M_i}\right)^{\frac{1}{2}} \tag{4}$$

since $J_e = J_i$ thus equating (2) and (4) gives

$$\Delta V = \frac{kT_e}{2e} \ln \left(\frac{\pi}{2} \frac{M_e}{M_i}\right) \tag{5}$$

In both formulas (3) and (5) the potential between the floating surface and the plasma is ΔV.

Vossen has quoted for an O_2 plasma at 300 mtorr pressure a floating potential V_f of -10 volts to ground and a plasma potential V_p to ground of +355 volts, thus ΔV is 365 volts. Assuming the chief ion present is O^+, and therefore $M_e/M_i = 3.4 \times 10^{-5}$ and that Ti does not exceed 400 K, then electron temperatures in the O_2 plasma in Vossen's system can be calculated using equations (3) and (5). The results from equation (3) = 37 eV and from equation (5) = 74 eV.

CHAPTER EIGHT

THE RF VOLTAGE/CURRENT CHARACTERISTICS AND RELATED DC NEGATIVE BIAS PROPERTIES OF AN ELECTROTECH FLAT BED PLANAR PLASMA ETCHER

8.1 Introduction

The aims and purpose of studying the ionization conditions arising in production type planar reactors has been discussed in 7.1 of the previous chapter. This chapter reports on the measurement of the rf voltage and current characteristics and the negative dc bias development at the insulated electrode facing the work surfaces when inert and active gases are present.

8.2 Experimental Work

8.2.1 Apparatus

The system used was an Electrotech 317 reactor with a 380 kHz rf generator capable of delivering 6 kW of power, as shown in fig. 7.5 in the previous chapter.

A simplified circuit diagram of the rf generator is shown in fig. 8.1 and a detailed description of the circuit is also given here because of the relation between the r.f. supply and the discharge characteristics.

The active device used was an ITT power triode with a water cooled anode (3 RM 1896). A three phase transformer was used to provide the high tension voltage (4.5 kV) and its output was full wave rectified and partly smoothed by an RC network. The output power from the anode was obtained via a capacitor C_1 (5000 pf) which blocked the dc

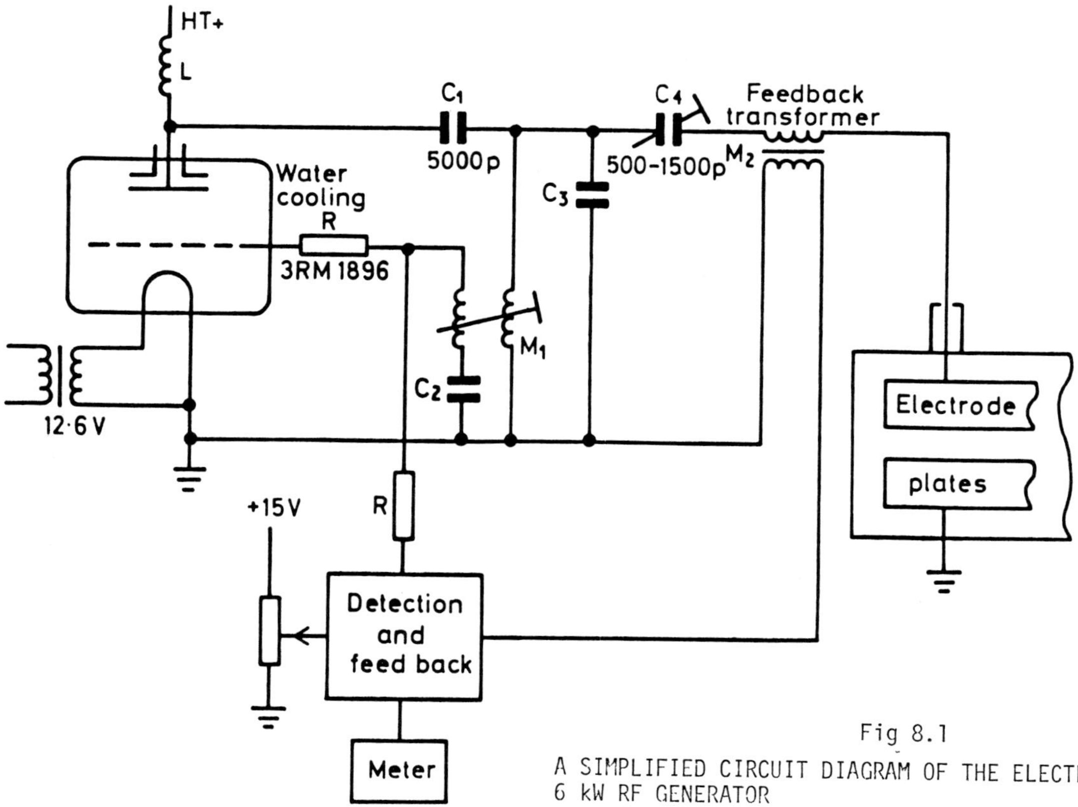

Fig 8.1

A SIMPLIFIED CIRCUIT DIAGRAM OF THE ELECTROTECH 0.38 MHz 6 kW RF GENERATOR

voltage on the anode. The tuned transformer M1 applied the required feedback from the anode to the grid to obtain oscillation. The rf power was applied to the top electrode via an adjustable blocking capacitor C_4 (500-1500 pf); this gave charge storage and enabled a negative dc bias to develop on the top electrode. The current transformer M2 in the connection to the top electrode measured the rf current flowing in the plasma. The output of the transformer M2 was rectified and smoothed, the dc output produced thus being proportional to the rf current. A DVM was used to measure this voltage and hence gave a direct readout of the rf current. Stabilization was obtained by comparing this voltage with a reference voltage using a difference amplifier, and the output of this amplifier was used to apply a dc negative feedback signal to the triode grid.

The planar reactor apparatus, apart from the addition of a potential divider was the same apparatus as is described in section 7.3.1.

The potential divider was connected from the top electrode to ground giving a 1000:1 reduction for monitoring the rf and dc voltages as shown in fig. 8.2. The voltage output from the potential divider was measured by a Tektronix 549 Storage oscilloscope. The oscilloscope was connected to the potential divider using a short co-axial lead to minimise stray capacitance. The negative dc bias potential to ground developed at the top electrode when connected to the blocking capacitor (C_4-1500 pf), was measured with an electrometer (Level TM9B). A choke (1 mH) was used to reduce the rf voltage to a negligible amount at the input terminals of the electrometer.

8.2.2 Negative bias potential to ground

Measurements were made under various conditions of r.f. current and pressure for each gas used. The relationship between dc negat-

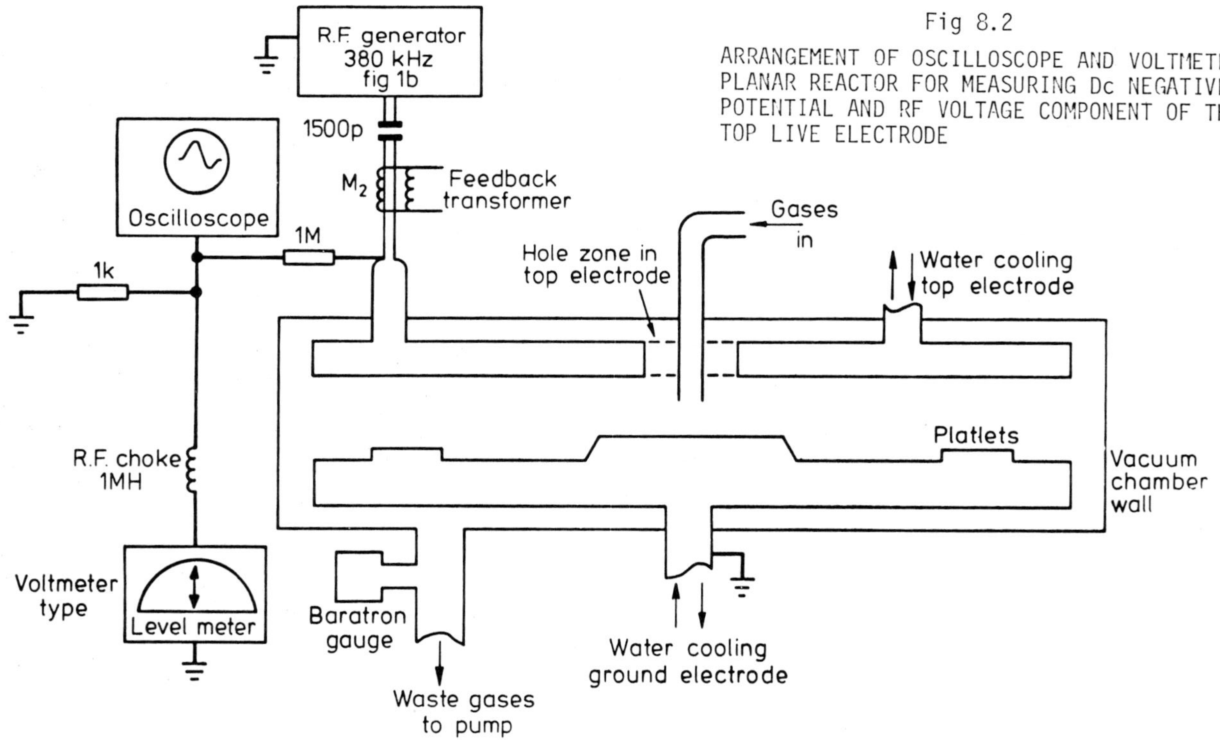

Fig 8.2

ARRANGEMENT OF OSCILLOSCOPE AND VOLTMETER TO PLANAR REACTOR FOR MEASURING Dc NEGATIVE BIAS POTENTIAL AND RF VOLTAGE COMPONENT OF THE TOP LIVE ELECTRODE

ive bias voltage and rf input current supplied from the rf generator, at a reactor pressure of 300 mtorr is shown in fig. 8.3 for Ar, O_2 and CCl_4 plasmas. A plot of dc negative bias voltage against pressure, at a constant input current supplied from the rf generator of 2 Amps, is displayed in fig. 8.4. This shows a decrease in the dc negative bias voltage with increasing pressure for oxygen and argon plasmas. However,

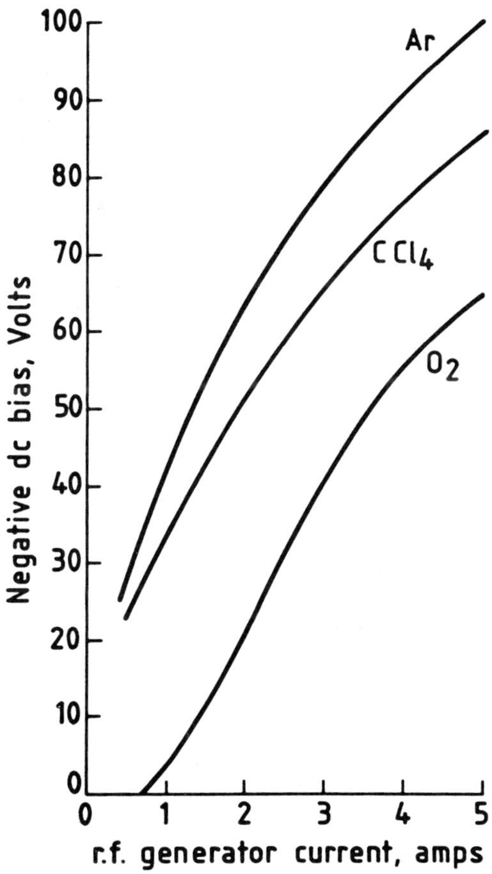

FIG. 8.3

DC NEGATIVE BIAS POTENTIAL OF THE TOP ELECTRODE AT 0.3 torr PRESSURE Vs VARIOUS CURRENT LEVELS FROM THE r.f. GENERATOR (rms)

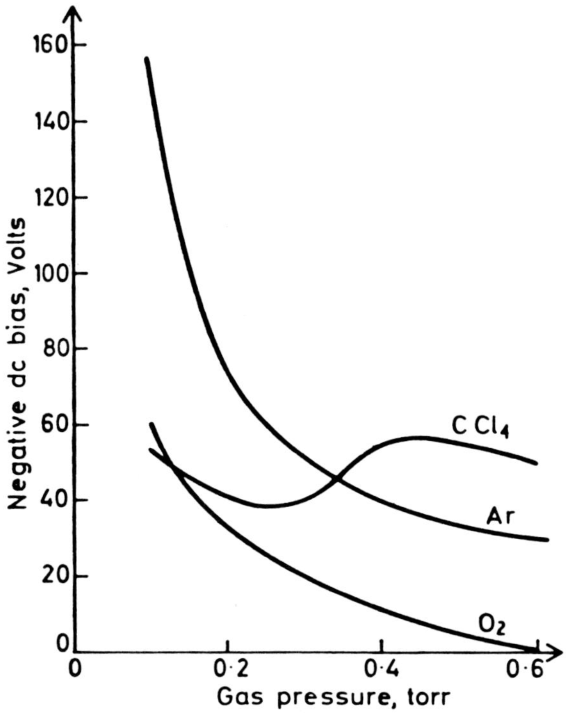

Fig 8.4

DC NEGATIVE BIAS POTENTIALS OF TOP ELECTRODE AT VARIOUS GAS PRESSURES. A CONSTANT CURRENT OF 2 AMPS (rms) SUPPLIED FROM THE RF GENERATOR

for carbon tetrachloride plasma there is a 'hump' between 300 mtorr and 600 mtorr pressure. This 'hump' coincides with a rise in the rf voltage for the carbon tetrachloride plasma in the 300 to 600 mtorr region and is discussed in the next section. Fig. 8.5 is a plot of dc negative bias voltage against pressure at a constant power of 500 watts, derived from figs. 8.3 and 8.4.

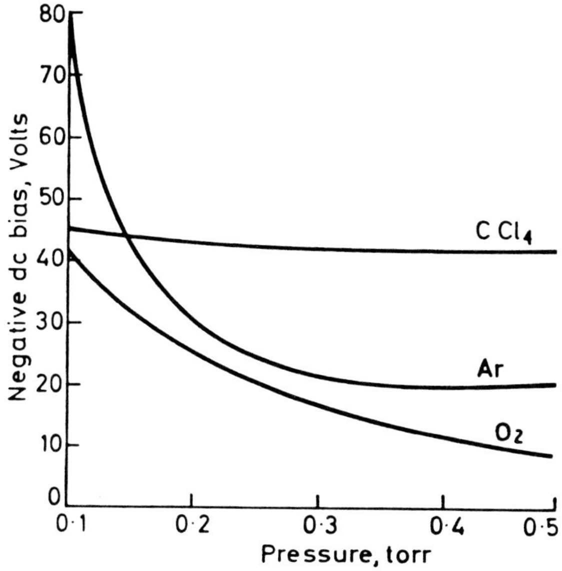

Fig 8.5

DC NEGATIVE BIAS POTENTIAL OF TOP ELECTRODE AT CONSTANT POWER FROM THE RF GENERATOR 500 WATTS (ESTIMATED) Vs GAS PRESSURE

8.2.3 RF voltage waveform and dependence on rf current

The oscilloscope shown in fig. 8.2 was used to display the rf potential appearing at the output of the potential divider. The rf voltage displays were recorded on Polaroid film and the peak voltages measured, from which the rms voltages were estimated. Measurements of r.f. voltages appearing on the top electrode were made under various conditions of rf current and gas pressure.

Figs.8.6 and 8.7 are photographs for a carbon tetrachloride plasma at 500 mtorr pressure, 1 and 5 Amps respectively, revealing distortion in the r.f. potential wave form at higher rf currents. Figs. 8.8 and 8.9 are photographs for an argon plasma at a constant r.f. current of 2 Amps, 100 mtorr and 500 mtorr pressure respectively, showing that there is a small ripple at the peak of the r.f. potential wave-

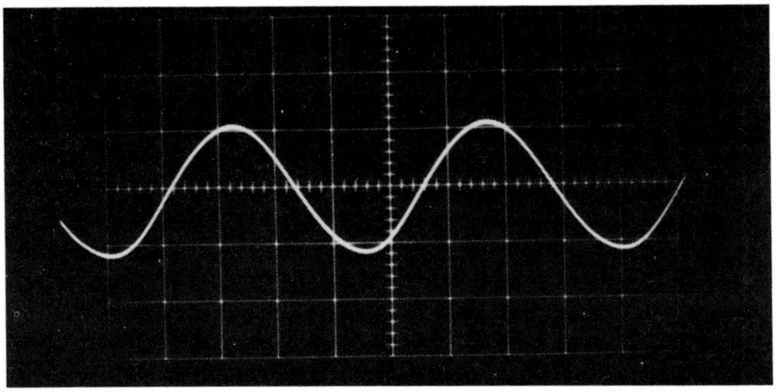

Fig. 8.6. CCl_4 plasma, 0.5 torr, 1 amp rms M_2, 500 V/D.

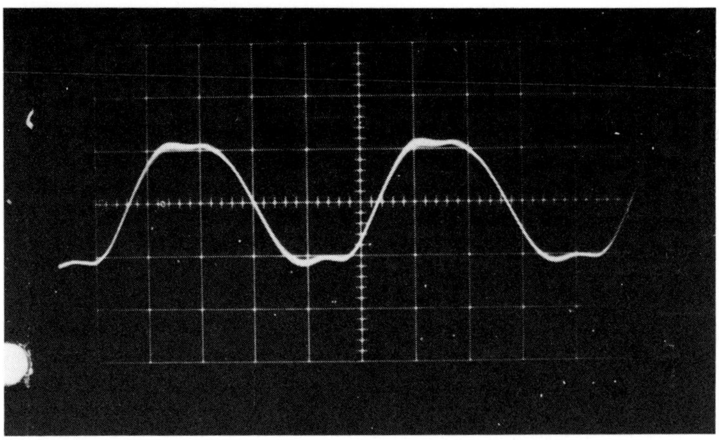

Fig 8.7. CCl_4 plasma, 0.5 torr, 5 amps rms M_2, 500 V/D.

form which is difficult to detect at higher pressures. The ripples were far more pronounced in Ar-plasma than in O_2 or CCl_4 plasmas.

Figs. 8.10 and 8.11 are plots of applied rf voltage (rms values) against rf current (rms values) supplied by the rf generator at 100 mtorr and 300 mtorr pressure. Carbon tetrachloride has a non-linear

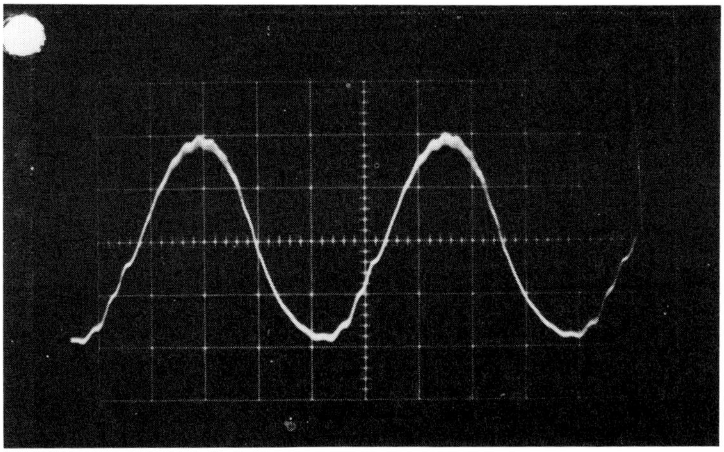

Fig. 8.8. Ar plasma, 0.1 torr, 2 amps rms M_2, 200 V/D.

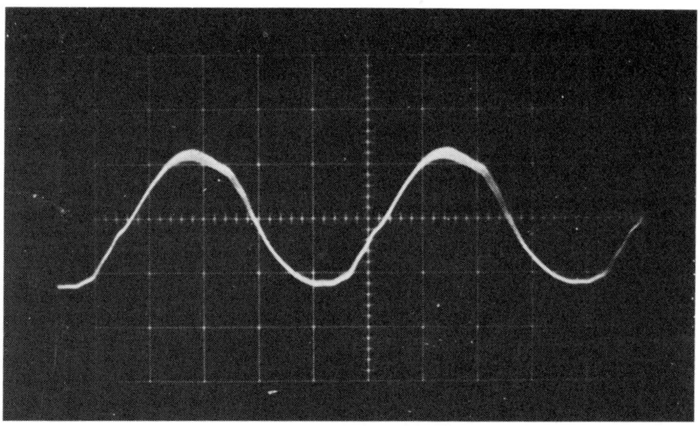

Fig. 8.9. Ar plasma, 0.5 torr, 2 amps rms M_2, 200 V/D.

characteristic at 300 mtorr pressure in comparison with Ar or O_2 plasmas at the same pressure, which are almost linear. Fig. 8.12 is a graph of rf voltage against gas pressure at a constant rf current of 2 Amps.

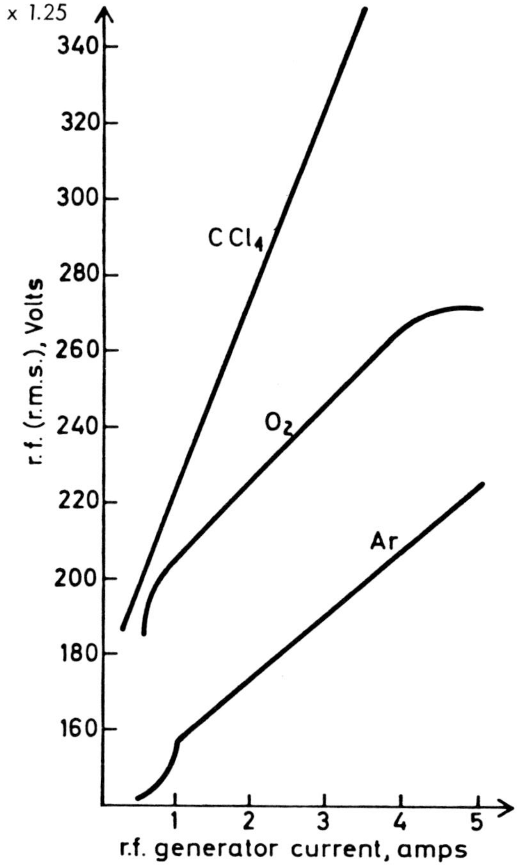

Fig 8.10

CCl_4, O_2, AND Ar PLASMA AT 0.1 torr PRESSURE rf rms VOLTS Vs CURRENT (rms) SUPPLIED FROM GENERATOR

FIG. 8.11

CCl_4, O_2 AND Ar PLASMAS AT 0.3 torr PRESSURE r.f. (rms) VOLTS Vs CURRENT (rms) SUPPLIED FROM GENERATOR

FIG. 8.12

RF VOLTAGE TO TOP ELECTRODE (rms) Vs GAS PRESSURE AT A CONSTANT CURRENT OF 2 AMPS (rms)

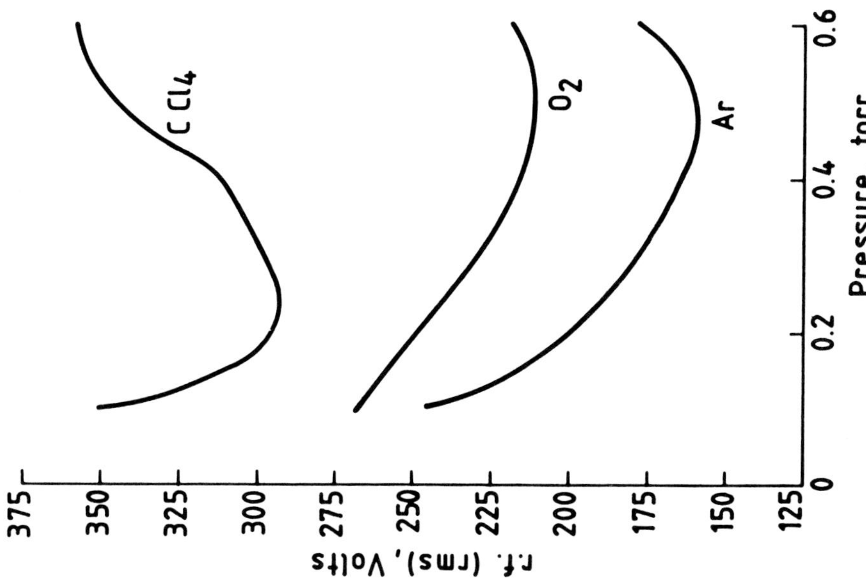
r.f. (rms), Volts
375
350
325
300
275
250
225
200
175
150
125
0
0.2
0.4
0.6
Pressure, torr
C Cl4
O2
Ar

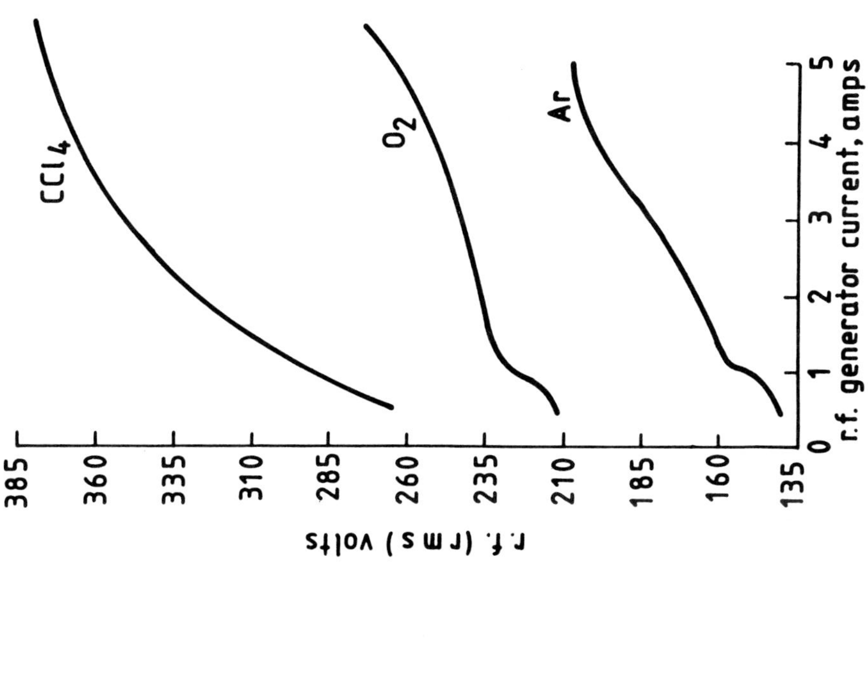
r.f. (r m s) volts
385
360
335
310
285
260
235
210
185
160
135
0
1
2
3
4
5
r.f. generator current, amps
CCl4
O2
Ar

8.2.4 The measurement of the rf current through a dummy ground electrode

In this experiment a dummy electrode was used to simulate the current flow through the ground electrode. The disks and holder which support the wafers during processing were removed from the ground electrode which was covered with a Pyrex glass disk. An aluminium electrode of the same diameter as the ground electrode rested on the Pyrex Sheet. A wire connection with ceramic sleeving was made to the dummy aluminium electrode and connected to ground via a vacuum electrical feedthrough to a 1Ω resistor outside the chamber, as shown in fig. 8.13.

The r.f. potential appearing across the 1Ω resistor (directly proportional to the current flowing in the dummy electrode) was displayed on the oscilloscope. The current flowing in the dummy electrode circuit was recorded using Polaroid film for various conditions of current supplied by the rf generator and gas pressure.

Figs. 8.14 and 8.15 are photographs for an Ar-plasma at a constant current from the rf generator of 2 Amps, at 100 mtorr and 300 mtorr gas pressure respectively. They show a small oscillation at the peak of the rf cycle when the rf top electrode is positive to ground, but there is no oscillation on the other half of the cycle when the electrode is negative to ground. These oscillations appear with O_2 and Ar plasmas under all conditions but are less evident at higher gas pressures.

Carbon tetrachloride plasmas had a different characteristic for their r.f. current waveform, as shown in fig. 8.16 which is a photograph of a CCl_4 plasma operated at an rf current of 2 Amps and at 300 mtorr gas pressure. Broad spikes can be see at the peaks on both halves of the rf cycle and these start to occur above 250 mtorr gas pressure.

Fig 8.13

MEASUREMENT OF r.f. CURRENT CHARACTERISTICS TO DUMMY GROUND ELECTRODE

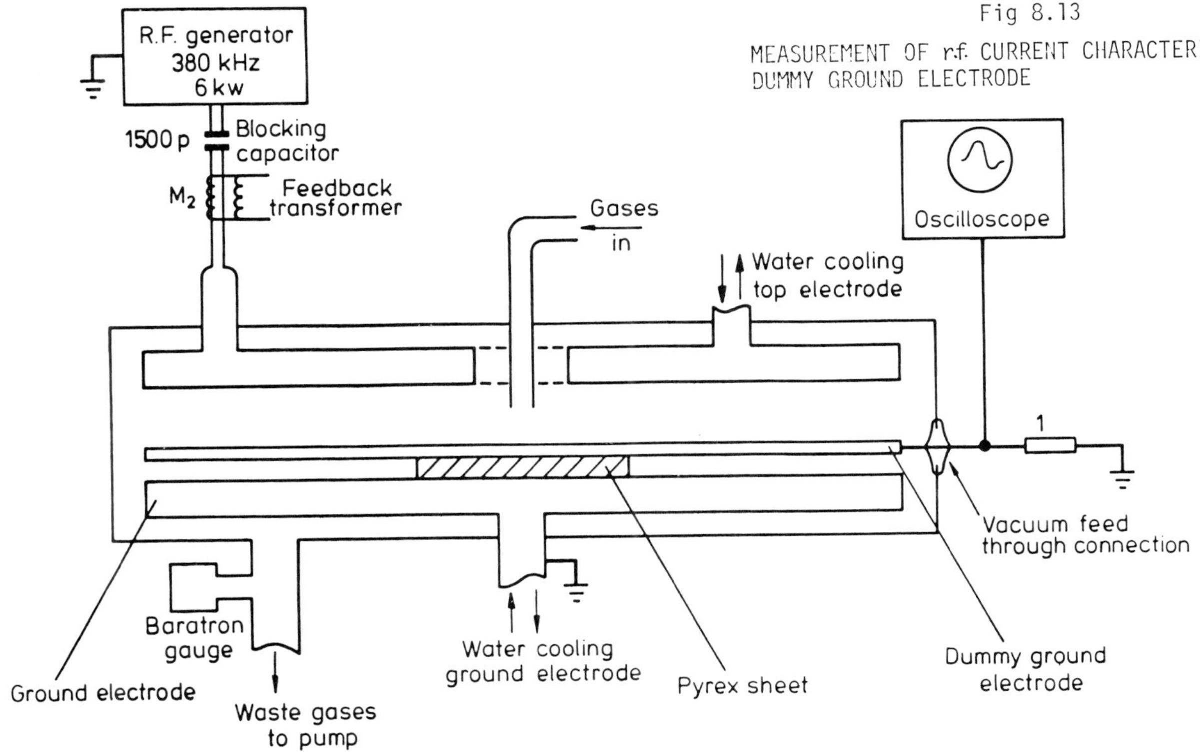

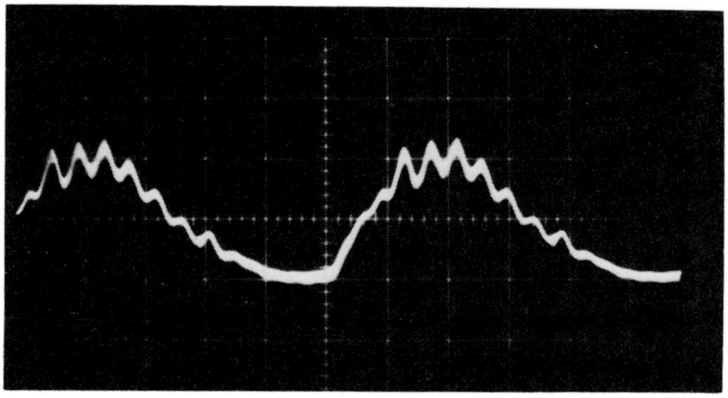

Fig. 8.14. Ar plasma, 0.1 torr, 2 amps rms M_2, 2 A/D.

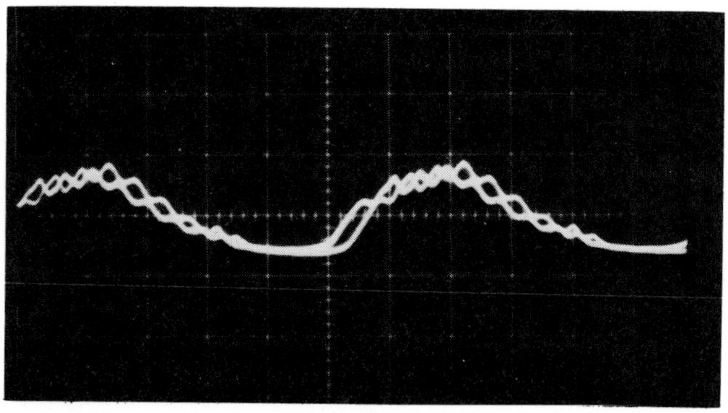

Fig. 8.15. Ar plasma, 0.3 torr, 2 amps rms M_2, 2 A/D.

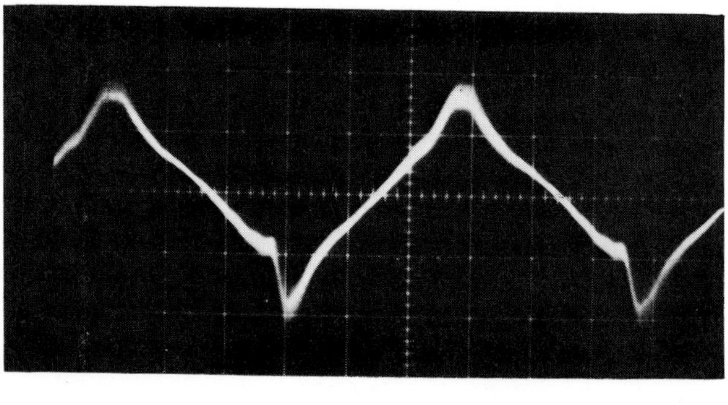

Fig. 8.16. CCl_4 plasma, 0.3 torr, 2 amps rms M_2, 2 A/D.

8.2.5 RF current and voltage display for phase comparison

The applied rf potential and rf current flowing through the dummy ground electrode were measured simultaneously as shown in fig. 8.17. The buffer amplifier type 517 was used for the rf voltage measurement and this was situated close to the potential divider. The buffer amplifier had a high input impedance and a low output impedance. This was incorporated in the circuit to minimise the effect of stray lead capacitance. A coaxial cable was used to connect the ground dummy electrode to the second channel of the oscilloscope, enabling the rf voltage and the rf current through the dummy electrode to be displayed at the same time.

The results of these experiments were recorded on Polaroid film for each gas and condition used, giving a record of the rf voltage and rf current, as shown by fig. 8.18 for a CCl_4 plasma at an rf current 1.5 Amps (M2), 300 mtorr pressure. Fig. 8.19 is a photograph for an Ar-plasma 2.0 Amps (rf M2), 300 mtorr pressure. The current passing through the 1Ω resistor and displayed on the oscilloscope is slightly lower compared with the meter measuring rf current via M2. Figs. 8.18 and 8.19 show that there is a small phase difference between the voltage and current waveforms.

8.3 Discussion

8.3.1 Voltages and the symmetrical system

The experiments have shown that a dc negative bias voltage with respect to ground is produced at the rf electrode, facing the grounded work holder. As the ground to plasma potential is not yet known, it is not possible to determine the potential between the top rf electrode and the plasma. However, the maximum negative bias to the plasma could not exceed the peak r.f. potential responsible for the electron extract-

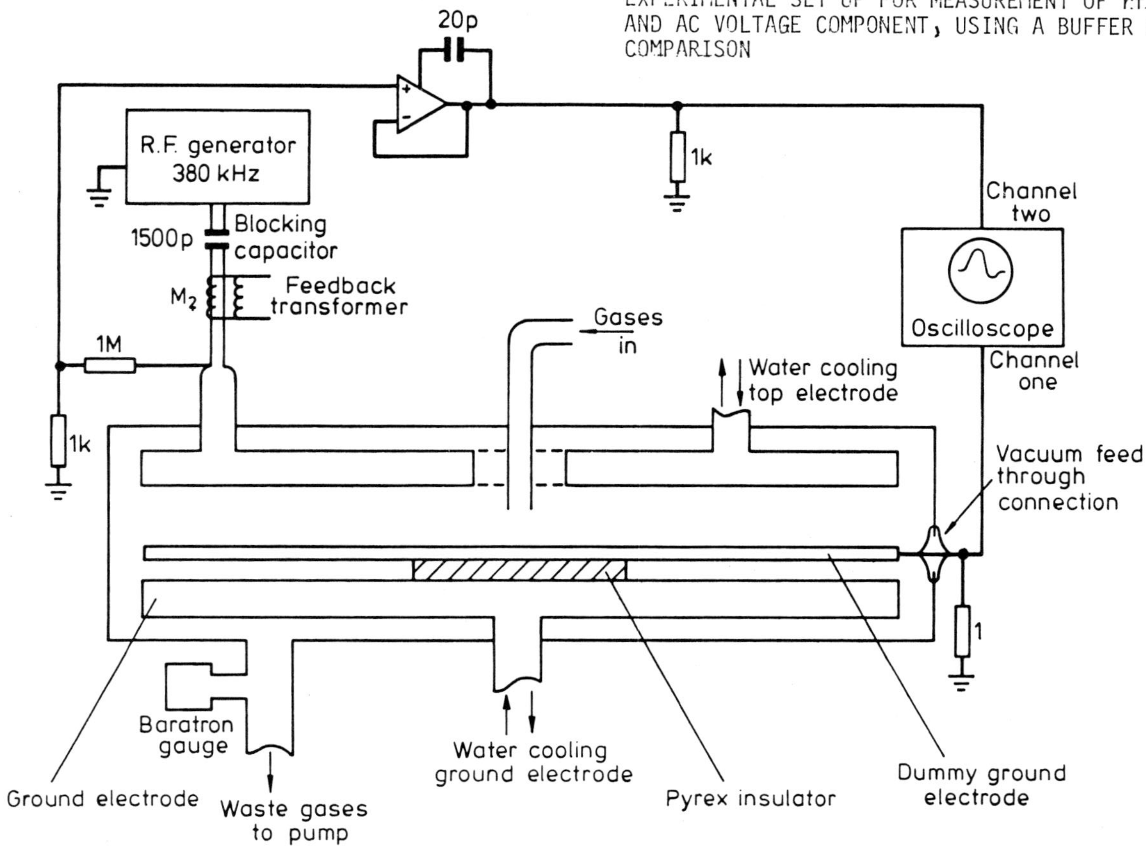

Fig 8.17

EXPERIMENTAL SET UP FOR MEASUREMENT OF r.f. CURRENT TO GROUND AND AC VOLTAGE COMPONENT, USING A BUFFER AMP FOR PHASE COMPARISON

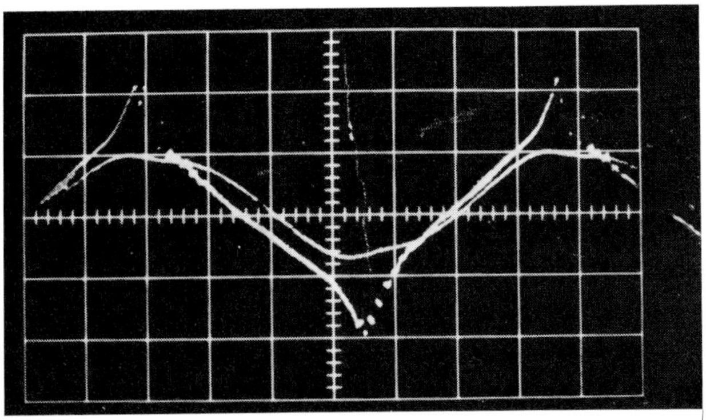

Fig 8.18 CCl_4 Plasma 0.3 torr 1.5 amps rms M2 500 V/D 1 A/D

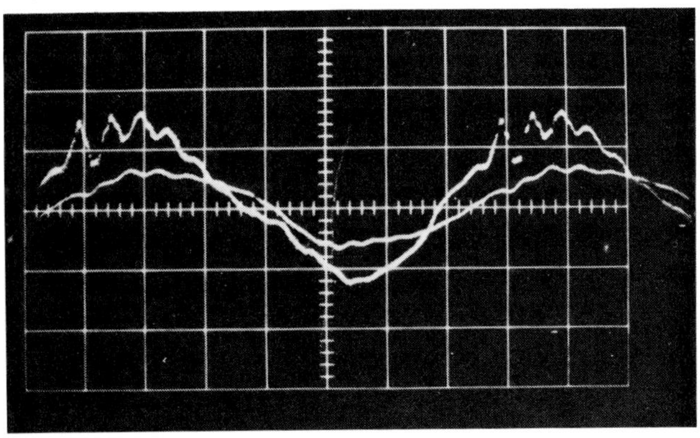

Fig 8.19 Ar Plasma 0.3 torr 2 amps rms M2 500 V/D 1 A/D

ion from the discharge electrode and would only be appreciable under conditions of negligible ion flow. Thus the bias must always be < -500V which corresponds to the peak r.f. value used.

The existence of a ground to plasma potential is shown by the development of a ground sheath approximately 2 mm depth corresponding to a similar sheath at the top electrode. Hence the negative dc bias voltage with respect to ground measured here may only represent a small difference between the two sheaths. In a similar system, Vossen (14) has measured both ground and top electrode potential with respect to the plasma using Langmuir probes. His results show a small difference between the two sheaths giving a negative dc bias potential with respect to ground of -30 Volts for an oxygen plasma at 0.1 torr gas pressure. The bias obtained in the studies here was ∿ -50 Volts for oxygen at 0.1 torr pressure and 2 Amps rms r.f. current. Naturally, for plasma etching and ion bombardment enhancement of etching, the ground sheath conditions are more important than the small differences between the two sheaths.

As discussed in the previous chapter, all attempts to use probes to determine plasma potentials have proved difficult in the present system employed because of the large charge densities attained under the ionization conditions existing at the high gas pressure used (300 mtorr). Operating a probe from the ion saturation region to the electron collection zone resulted in probe currents of the order of several milliamps with the outcome that the probe overheated even at a small voltage of +10 volts with respect to ground. However, using the double probe method electron temperatures of 8 eV and ion densities up to 10^{18} ions m^{-3} have been measured for an Ar plasma at 0.3 torr pressure and 2 Amps rms. These conditions represent one single charged ion per 10^4 neutral atoms. Vossen (14) does not give the r.f. current or power condition used with his plasma etcher and thus he may not be operating at such high ion/electron densities.

8.3.2 Influence of bias on etch rate

An experiment was conducted to determine the etch rate of aluminium and silicon at the ground electrode and top electrode. Aluminium films were etched marginally faster at the top electrode.

However the difference was very small and it was difficult to determine accurately due to non-uniform etching across the wafers at the top electrode. Silicon under the same conditions (300 mtorr gas pressure 2 Amps rms) was etched almost 50% faster at the top electrode. The higher etching rate for silicon at the top electrode could arise from a greater dependence on ion impact enhanced etching than for aluminium. The dependence on ion impact etching for silicon in this system is high, which was born out by the lack of lateral etching in the Si films. Aluminium suffers from appreciable lateral etching producing an anistropy for this system of $\sim$ 2.5 when the sample was etched on the ground electrode.

The system conditions were similar for the two etching experiments. The sheath bias (ΔV) was 40 Volts greater at the top electrode under the same etching conditions (CCl_4 plasma at 300 mtorr pressure, 2 Amps rms r.f. current) for both the Al and Si etched samples. The difference in the sheath potentials can arise from a difference in the ion current densities (N_i) in the two sheaths. If they were the same, the system would be symmetrical, i.e. balanced, and no nett bias to ground would be measured. However, if etching is enhanced by bombardment of ions accelerated in an adjacent sheath, then one would expect a higher etch rate at the top electrode than the ground holder because of its greater power density; the etch rate $\varepsilon(JV_i)$, where J is the current density and V_i is the ion energy. Although the ratios of etching rates for Si and Al differ, we can conclude that the top electrode is likely to give the highest etch rate.

Further experiments are in progress to determine the ground to plasma potential (67). Recently Taillet (154) has calculated, for typical operating conditions, the sheath potential drop likely to be developed in a plasma reactor used for wafer etching. His model was based on a symmetrical electrode system and the calculation was made to determine the sheath width in the wafer region. However, the results presented here indicate that in at least one kind of practical system there is a small voltage difference between the top electrode sheath and ground sheath. Although the system does appear to be symmetrical, the area of the ground surfaces in contact with the plasma must be slightly larger than the top r.f. electrode area (ref 155). Taillet's calculation appears to be based on a symmetrical system and such calculations do not appear feasible until we can either determine the plasma potential or the areas of the electrodes active in the discharge excitation and their related current density distribution.

Although plasma potentials for the system discussed here could not be directly determined, the r.f. voltages applied to the electrodes are known and from previous discussions plasma potentials must always be the most positive potential in the system to limit the loss of electrons from the plasma.

Fig. 8.2 shows the electrode system with a blocking capacitor connected between the r.f. generator and the top electrode. The bottom electrode is grounded. With a blocking capacitor in the system, charge storage can occur resulting in a d.c. bias to ground developed at the top (r.f.) electrode). At the frequency used, 380 kHz, a small d.c. negative bias to ground was observed. This is in contrast with another commonly used frequency 13.5 MHz, where other workers (66) have observed a negative d.c. bias to ground comparable to the r.f. peak voltage.

Examine next the charge flow through the r.f. cycle. When an r.f. electrode with capacitive coupling is at its peak potential on the positive part of the cycle, the plasma to r.f. electrode potential, (V_p - $V_{r.f.}$) is at its lowest and may only be a few volts. This allows the more energetic electrons from the plasma to surmount the potential barrier and reach the live r.f. electrode, where charge storage can result in a dc negative bias. Once past the peak, the plasma to ground potential decays at a rate governed by the ion flow from the plasma until the plasma to ground potential is low enough for the more energetic electrons to surmount the potential barrier of the ground electrode sheath and restore charge loss. At 380 kHz, the rate of fall of the plasma potential is rapid enough for it to roughly follow the drop in the r.f. electrode potential during the positive cycle change (132). However the plasma potential remains a few volts above the r.f. electrode potential during the positive cycle and a few volts above the ground potential during the negative cycle and hence is always the most positive potential of the system. A simple diagram of these changes in potential and charge flow is shown in fig. 8.20.

Assuming that on the positive part of the r.f. cycle $V_p - V_{r.f.}$ is negligible, then the voltage drop across the electrode sheaths can be estimated from the measured r.f. applied potential and dc negative bias at the r.f. electrode. For a CCl_4 plasma at 300 mtorr, 2 Amps rms the r.f. peak-to-peak voltage is about 930 volts and the dc negative bias is $\sim$ 40 volts with respect to ground. Thus the maximum voltage drop across, (i) the ground sheath (V_g) is 425 volts and (ii) the r.f. electrode sheath (V_t) is $\sim$ 505 volts. If, however, $V_p - V_{r.f.}$ is not negligible and assumes a value of say 30 volts then V_g = 455 and V_t = 535 volts.

The small d.c. negative bias to ground developed on the live electrode at 380 kHz excitation frequency is believed to be because the effective ground area is slightly greater than that of the live r.f. electrode area. This result can be derived from the model of charge flow (129) once the condition $\oint Idt = 0$ to all surfaces is applied at steady state conditions. When the plasma is operated in a steady state

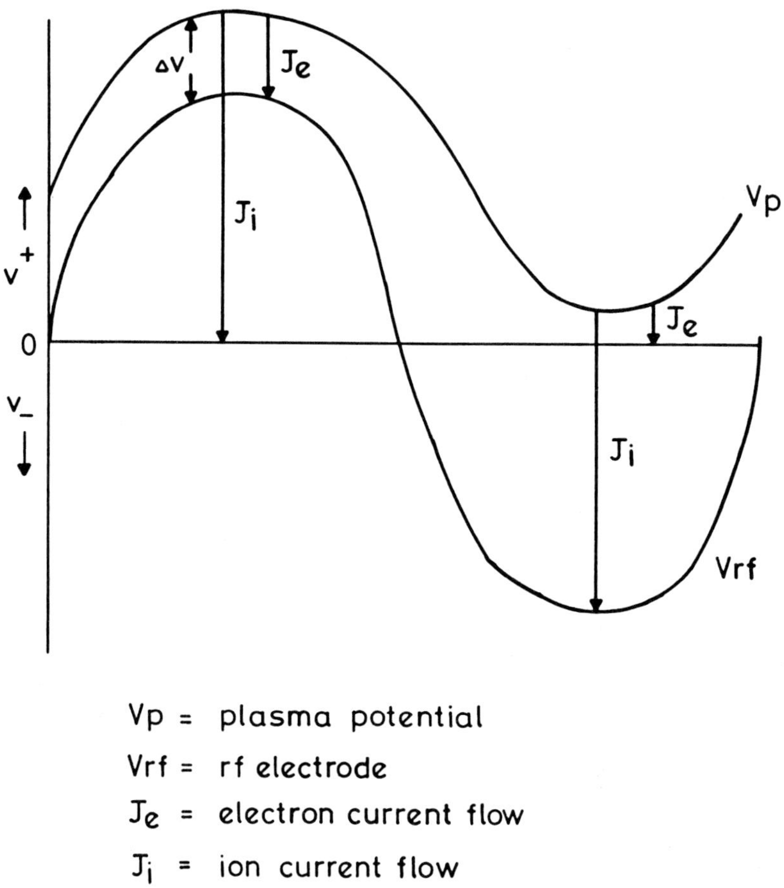

FIG 8,20

CHARGE FLOW DURING rf CYCLE

(i.e. the potential of each part of the system returns to the same value at the completion of each cycle) then the net charge flow must equal zero over a cycle i.e. of $\oint Idt = 0$. If one electrode is grounded and its effective area is larger than that of the r.f. electrode then the ion density to the ground electrode will be lower than that to the r.f. electrode. Consequently, the sheath potential will be less at the ground electrode and the plasma will be less positive to ground than to the r.f. electrode as shown in figs. 8.21 and 8.22. Under these conditions,a negative bias will exist between the two electrodes which is the difference between their sheath potentials.

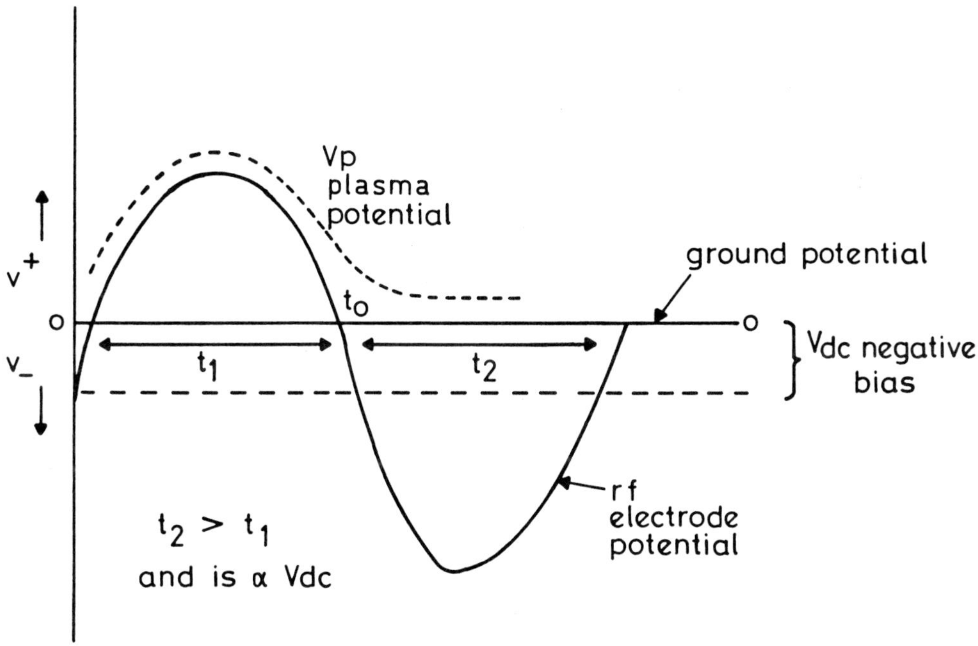

FIG 8,21

DC NEGATIVE BIAS ON LIVE ELECTRODE

If there is an increase in effective ground area of a system such as walls to the vessel which are grounded then the current to the actual ground electrode itself will be reduced. However, the currents to the total ground area and live electrode area will be equal.

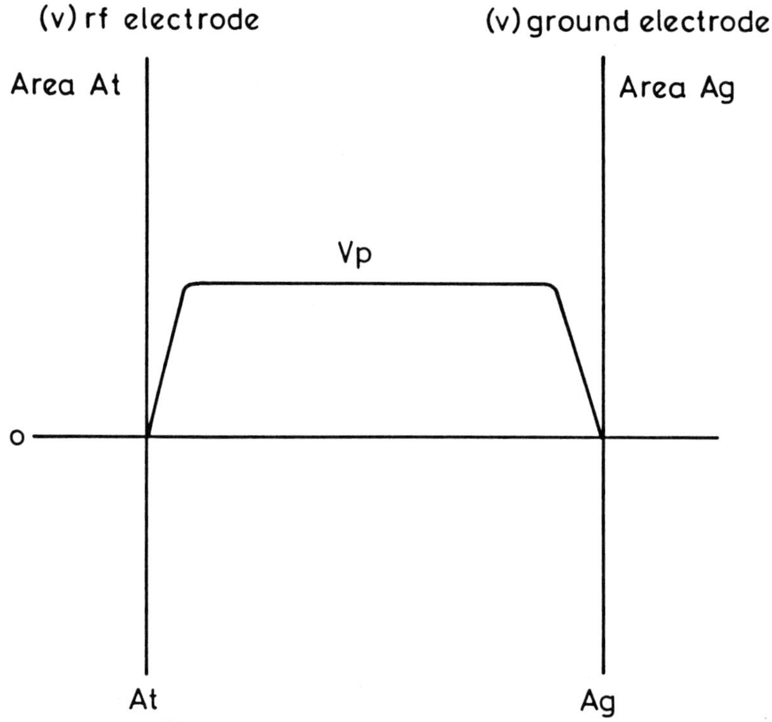

Symmetrical Electrode System
At = Ag

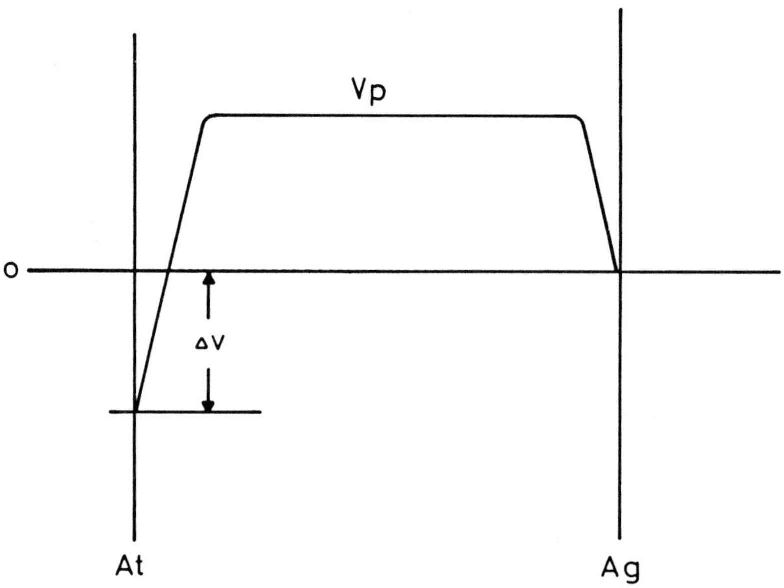

FIG 8.22
Asymmetrical Electrode System
Ag > At

The rms currents to the dummy ground electrode decrease as the pressure decreases which coincides with an increase in the observed negative bias of the r.f. electrode to ground shown in fig. 8.4. The r.f. current waveform flowing to the dummy ground electrode is shown in fig. 8.14. The rms value was estimated to be 1.4 Amps, with an r.f. current of 2 Amps rms. The Ar plasma was maintained at a pressure of 100 mtorr. The dc negative bias to ground at these conditions was about 156 volts. As the dummy electrode shows a smaller current flowing, this indicates we are dealing with an asymmetrical system; because there are alternative plasma to ground paths. The increase in dc negative bias on the rf electrode observed when pressure was decreased for the O_2 and Ar plasmas as shown in fig. 8.4 is probably because the diffusion path of the ion has increased. This will enable the plasma to contact other surfaces which are most likely at ground potential. The CCl_4 plasma does not have the same characteristic since the applied rf potential reduced to a minimum at about 250 mtorr.

8.3.3 The voltage plots

Fig. 8.10 shows the approximate linear plot of rms voltage against current supply current for Ar, O_2 and CCl_4 plasmas maintained at 100 mtorr pressure. However, fig. 8.11 shows some curvature of rms voltage plotted against current at 300 mtorr pressure. This curvature or non-linearity is probably because the charge distribution in the system is becoming non-uniform which has also been observed in chapter 7, where the T_e values decrease with increasing pressure.

In fig. 8.11 the CCl_4-plot shows more curvature than the Ar and the O_2 plots which still have some linearity. This is probably due to the shorter mean free path of the secondary electrons in the CCl_4 plasma, thus causing more non uniform charge distribution. This could also coincide with the much larger change in T_e values from 100-300 mtorr pressure in CCl_4-plasma than in the Ar plasma for the same change, e.g. CCl_4 100 mtorr T_e = 16 eV and 300 mtorr T_e = 14 eV for Ar 100 mtorr T_e = 8.7 eV and 300 mtorr T_e = 8.0 eV.

8.3.4 The multiplication process

Fig. 8.16 shows a photograph of a CCl_4 plasma at 300 mtorr showing the 'spikes' in the current trace at each peak of the cycle. The spikes occur when the applied potential was near its maximum, i.e. when the electric field across the space charge sheath was near its maximum. The observed spikes occur at a pressure of 250 mtorr and above for the CCl_4 plasma. This effect, however, has not been observed for Ar or O_2 plasmas in the system here. The spikes were probably due to a multiplication process within the space charge sheath when the electric field was high enough to accelerate the secondary electrons to ionization energies within a short distance λ_e. The electron mean free path, λ_e in a CCl_4 plasma is very much smaller than in Ar or O_2 plasmas and if the mean free path is smaller than the space charge width (d) then collision within the sheath can ionize neutral molecules which will cause a multiplication process as shown in fig. 8.23.

The lack of spikes for the Ar, O_2-plasmas are due to the longer electron mean free path and the smaller fields which do not contribute so readily to a multiplication process at these conditions.

Other workers, (133) and (156), using Cl_2 plasma at 300 mtorr (100 kHz) and H_2 plasma at 27 torr (210 kHz) have also observed similar spikes on the rf current waveform, at low frequency and high pressures. There are two possible reasons why multiplication is not likely to occur at higher frequencies; (1) the rf potentials are usually lower; thus the electric field is lower and (2) the current is reactive and out of phase with the voltage.

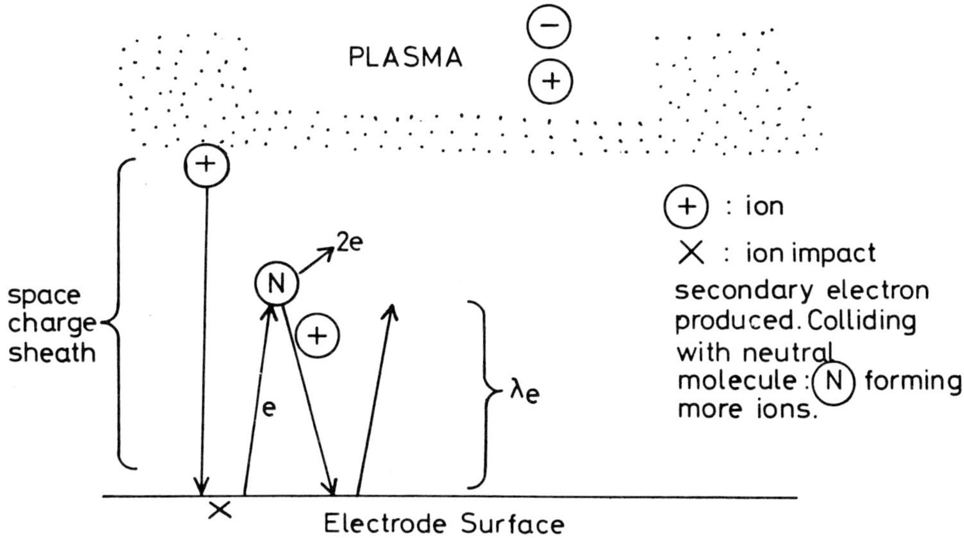

FIG 8.23
THE MULTIPLICATION PROCESS

It is possible to estimate the relation of λ_e/d for O_2, Ar and CCl_4 plasmas and predict the onset of multiplication processes as the pressure is increased. At low pressure, the ion mean free path is greater than the width of the space charge sheath and thus ion transit is effectively collision free. The solution of Poisson's equation:

$$\frac{d^2V}{dx^2} = - \frac{eN_e}{K_o} \tag{A}$$

where V = Voltage across sheath

x = distance across region of space charge

K_o = Permitivity of free space

N_e = Charge Density

e = Electron Charge

is the standard result for space charge limited current where:

$$d = 0.85 \left(\frac{k_o}{J}\right)^{\frac{1}{2}} \left(\frac{e}{m_i}\right)^{\frac{1}{4}} V^{\frac{3}{4}} \qquad \text{(B)}$$

where d = sheath thickness, m_i = ion mass and J is current density. Consider now higher pressures where there are ion collisions in the sheath. Assuming that the ion after a collision, in the sheath starts again from rest and undergoes uniform acceleration f, travelling distance x. It then impinges with the electrode surface with a velocity of impact equal to:

$$V = (2fx)^{\frac{1}{2}} \qquad (\text{from } V^2 = 2fx) \qquad \text{(C)}$$

The probability that the ion will have survived without a collision for the distance x is e^{-x/λ_i}, where λ_i is the ion mean free path ($\lambda_i \approx \lambda_{gas}$). The acceleration f of an ion is p = mf

$$\therefore \qquad f = \frac{eE}{m_i} \qquad \text{(D)}$$

where E is the electric field, the approximation being that it is constant over the last mean free path before collision with the electrode. The basic equations governing field, voltage and ion transit are (A), (B), (D) and J = N_e eV and thus if the voltage is known then the sheath thickness can be estimated from:-

$$d = 2.5 \left(\frac{k_o}{J}\right)^{2/5} \left(\frac{e\lambda_i}{2m_i}\right)^{1/5} V^{3/5}$$

for collision conditions with small values of λ_i. Assuming $2.5 \left(\frac{k_o}{J}\right)^{2/5}$ is a constant (B) estimated to be 4.93×10^{-5}, then

$$d = B\left(\frac{e\lambda_i}{2m_i}\right)^{1/5} V^{3/5}$$

If we consider the ground sheath, then $V^{3/5}$ can be estimated from the applied voltage data and the negative dc bias dara from figs. 8.12 and 8.4. Assuming the value $\frac{1}{2}\underline{V_{rms}} - V_{dc}$ to be the average potential existing across the sheath, d* can then be estimated. The electron mean free path λ_e can be estimated from the viscosity data of the gas considered.

$$\lambda_g = \frac{h}{\frac{1}{2}\rho u}$$

where $u = \sqrt{\frac{8kT}{\pi m}}$

$\lambda_e = \lambda_g \times 5.6$ (λ_g = mean free path of gas)

h = viscosity

ρ = density

u = velocity

m = mass of molecule of gas

Table 8.1 shows the data of d/λ_e for O_2, Ar and CCl_4 plasmas 2 Amps rms at various pressures and these are plotted in fig. 8.24.

Fig. 8.24 shows the larger d/λ_e values obtained for CCl_4 plasmas compared with O_2 and Ar-plasmas, all being at the same pressure and r.f. current input. When $\lambda_e < d$ the probability of an electron having an ionization collision in the dark space increases. Therefore CCl_4 plasmas with the larger values of d/λ_e are more prone to multiplication within the dark space than O_2 or Ar plasmas at the same pressure and current. In a CCl_4 plasma, multiplication, i.e. spikes, were observed in the r.f. current trace to the dummy electrode once above a pressure of 250 mtorr which corresponded to a d/λ_e value of ~ 3.5. Thus for significant multiplication to be observed a value for d/λ_e must be $\gtrsim 3.5$.

---

*d is an average dark space width over one cycle.

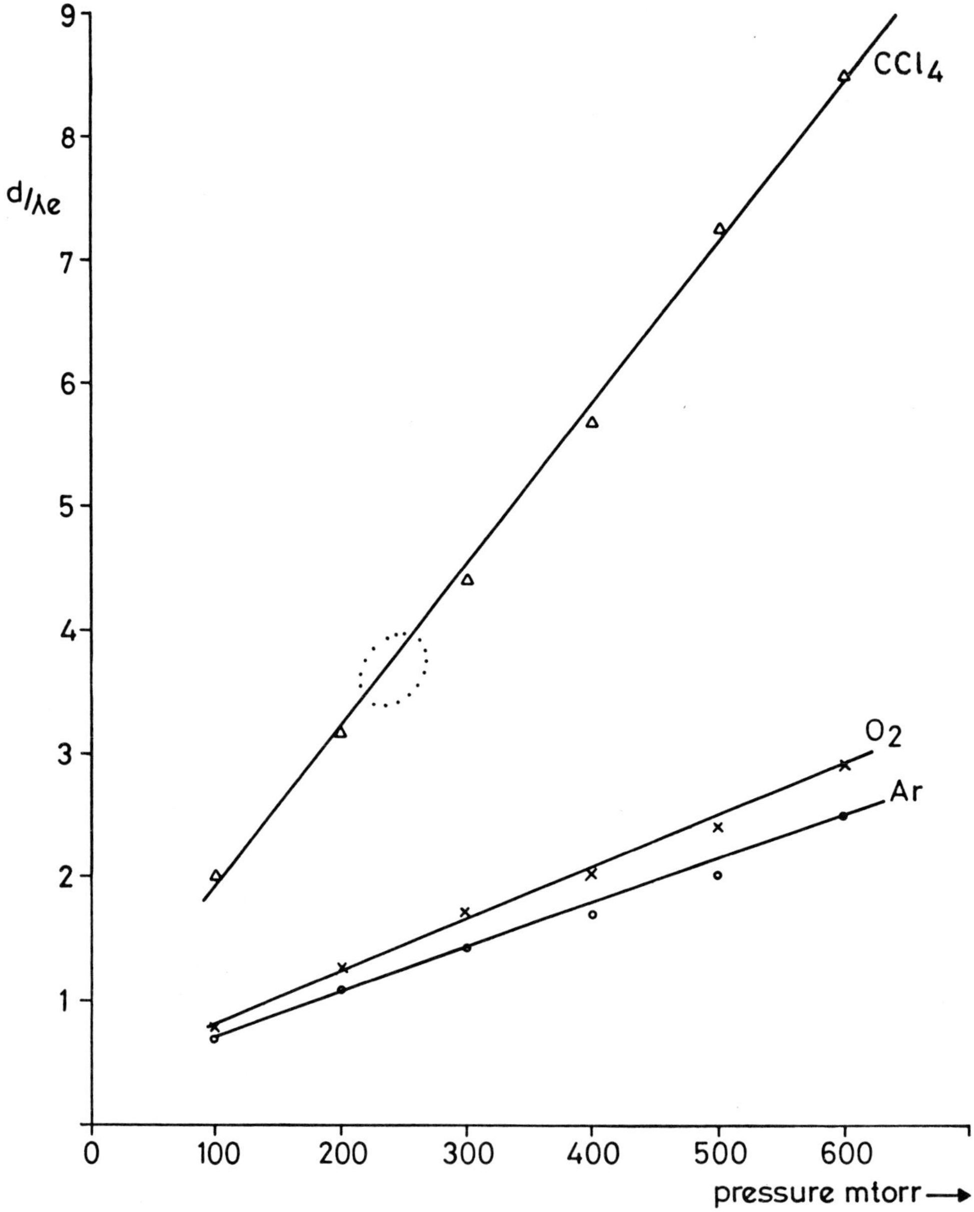

FIG 8.24

PLOT OF d/λ_e AGAINST PRESSURE. MULTIPLICATION OCCURING ~ 3·5 d/λ_e AT 2 Arms

Table 8.1

Gas and Pressure mtorr	Gas Mean-free Path μm	Sheath Thickness d μm	λ_e μm	$\frac{d}{\lambda_e}$ x10^{-1}	$\frac{d}{\lambda_e}$
Oxygen					
100	760	3320	4294	7.7x10^{-2}	0.77
200	380	2720	2147	0.1266	1.26
300	253	2430	1429	0.170	1.7
400	190	2220	1073.5	0.200	2.0
500	152	2070	858.8	0.240	2.4
600	126	2060	711.9	0.290	2.9
Argon					
100	730	2900	4124.5	7.0x10^{-2}	0.7
200	365	2260	2062.35	0.109	1.09
300	243	1970	1372.9	0.143	1.43
400	182	1770	1028.3	0.172	1.72
500	146	1650	824.9	0.200	2.0
600	121	1720	683.65	0.25	2.5
CCl_4					
100	190	2150	1073.5	0.200	2.0
200	95	1700	536.75	0.316	3.16
300	63	1570	355.95	0.441	4.41
400	47.5	1530	268.37	0.570	5.70
500	38	1560	214.7	0.726	7.26
600	31.6	1520	178.5	0.851	8.51

At 2Amp rms

Fig. 8.12 shows a decrease in the rms applied voltage with increase in pressure for all gases. However, when $d/\lambda_e > 3.5$, such as in the CCl_4 plasma there appears to be a rise in voltage because collisions are taking place between the secondary electrons and neutrals.

These collisions can reduce the electron energy, by inhibiting the electron acceleration to full sheath potential. Thus the applied voltage must increase to compensate and allow sufficient energy for ionization to be obtained between collisions.

The voltage is highest at the peaks of the r.f. cycle and at this point the electrons obtain sufficient energy to ionize neutral molecules within a distance much less than the sheath thickness. For Ar and O_2 gases, this only appears to start at higher pressures because λ_e is larger. Thus multiplication will only occur at above $\sim$ 600 mtorr. The rise in rms voltage for O_2 and Ar at high pressures in fig. 8.12 could be due to the non-uniform charge distribution.

As collisions occur at $d > \lambda_e$, there will also be a large generation of free radicals as their energy threshold for formation is lower than that of the ions. Thus free radical formation (multiplication) will also take place as the electric field and pressure increases. The multiplication process producing free radicals in a CCl_4 plasma could explain the observed increase in the etch rate of aluminium / 2% silicon films once above $\sim$ 250 mtorr pressure (67). Because of the large number of free radicals generated in the sheath in the vicinity of the etched sample with multiplication processes, poor etching anisotropy is obtained with aluminium/2% silicon.

8.3.5 Ion calculation and floating potential estimation

Using fig. 8.16, the rms value for the current collected by the dummy electrode for a CCl_4-plasma (maintained at 300 mtorr pressure and 2 Amps input current) has been estimated graphically from fig. 8.16 to be about 1.8 Amps rms (electrode diameter of 60 cm). The etch rate at the ground electrode in the system under these conditions is 1900Å per min for polysilicon and etches anisotropically. The density of polysilicon has been estimated to be 2.3 gm cm^{-3}, thus the amount removed per second, represents a volume of 3.16×10^{-7} cm^3 per sec of material removed and is a mass loss per second of 7.26×10^{-7} gram sec^{-1} per cm^2 area. Thus the number of atoms removed per cm^2 per second is $\frac{7.26 \times 10^{-7}}{28} \cdot N = 1.5 \times 10^{16}$ atoms $cm^{-2}sec^{-1}$, where 28 is the atomic weight of silicon and N is Avogadro's number, 6.02×10^{23} $mole^{-1}$.

The etch rate of polysilicon at 2 Amps rms, 300 mtorr is 1900Å/min and represents 1.5×10^{16} atoms of silicon leaving 1 cm^2 of silicon sample surface per second. It is instructive to compare this value with the charge arriving per second per cm^2 at the sample surface. From the positive half of the r.f. cycle to ground shown in fig. 8.16, the current to the dummy ground electrode has been estimated to be $\sim$1.8 Amps rms. This is a current density to the dummy electrode of 6.3×10^{-4} Amps cm^{-2} and represents an electron charge of 10^{16} per cm^2 per second. If the assumption is made that $q_e = q_i$, i.e. the positive (q_i) and negative (q_e) charge flow must balance, then this is equivalent to 10^{16} ions per cm^2 per second. This ion flow rate per cm^2 is close to the silicon etch rate of 1.5×10^{16} atoms per cm^2 per second.

This simple calculation makes many assumptions. One was that the current density of the dummy electrode is approximately equal to the original ground electrode. The change in geometry when incorporating the dummy electrode may reduce the current density arriving at the dummy electrode. As more grounded surfaces may be exposed to the plasma when incorporating the dummy electrode. Additional errors due to charge non-uniformity and multiplication processes make measurement of the average current density at the ground electrode difficult.

However, if the current density is correct, then the calculation indicates that the reaction between the CCl_4 plasma and polysilicon must rely on free radicals as well as ions since the etch rate exceeds the ion impact rate. The etching of polysilicon produces a profile with no measurable undercut indicating an ion enhanced reaction. It may be that the ion bombardment damages more than one Si-Si bond per impact allowing free radical reactions to proceed at the damaged sites.

If a floating body is introduced into the CCl_4 plasma, i.e. a body isolated from the ground electrode by an insulator, then the body will develop a negative potential V_f to the plasma. This will produce a potential barrier of $V_p - V_f$ (as an ion sheath) to reduce the energetic electron flow and maintain a balance between ion and electron flow arriving to the body. From the previous chapter, T_e has been estimated to be $\sim$ 14 eV for a CCl_4 plasma at 300 mtorr 2 Amps rms and from this it is possible to estimate $V_p - V_f$ by using the Podgorgi equation:

$$V_p - V_f = \frac{kT_e}{2e} \ln\left(\frac{M_i\ T_e}{m_e\ T_i}\right)$$

where V_p = plasma potential

V_f = floating potential

T_i = ion temperature (which is about $400^{o}C$)

m_e = electron mass

If M_i is assumed to be for CCl_3^+, then the potential difference $V_p - V_f$ from the equation is 35 volts. This is the estimated voltage drop across the sheath of a floating surface in a CCl_4 plasma. When wafers with polysilicon and aluminium films were floated in the system at 300 mtorr 2 Amps rms on petri dishes; etching did not proceed. This could be due to the voltage threshold not being reached for either removal of a surface oxide layer by sputtering or enhancement of chemical etching by ion impact.

8.3.6 Practical solutions

Plasmas examined had estimated ion densities of $\sim 10^{18}\,m^{-3}$ and could develop large r.f. electrode potentials above 1000 volts peak to peak for a CCl_4 plasma. The dc negative bias with respect to ground developed on the r.f. electrode i.e. the difference between the ground and r.f. electrode sheaths were small, between 40 and 60 volts for a CCl_4 plasma, 2 Amps. The floating potential in a CCl_4 plasma at 300 mtorr pressure 2 Amps rms has been estimated to be 35 volts negative to the plasma.

The electron temperature (T_e) for a CCl_4 plasma at 100 and 300 mtorr is 16 and 14 eV respectively. The fall in T_e with pressure rise coincides with a change from the linear I-V plot shown in fig. 8.10 at 100 mtorr to a more curved I-V plot shown in fig. 8.11 at 300 mtorr. The T_e and I-V plot changes between 100 and 300 mtorr pressure for O_2 and Ar plasmas are smaller. This may be because of the complexity of the CCl_4 plasma where the larger fragmentation and shorter electron mean free path results in non-uniform charge distribution.

The optimum conditions for etching aluminium/2% silicon on a SiO_2 substrate with a photoresist mask was a CCl_4 plasma at a pressure of 300 mtorr with a constant current of 2 Amps rms (67). These conditions produced the best selectivity between aluminium/2% silicon and photoresist mask layer with a reasonable aluminium/2% silicon etch rate. Fig. 8.12 shows that there is a drop in the r.f. applied potential to a CCl_4 plasma around the optimum etching conditions of 300 mtorr pressure. The improvement of selectivity at 300 mtorr is probably because of the large generation of free radicals from the multiplication process, increasing the aluminium/2% silicon etch rate. At the same time,the photoresist etch rate is almost constant from 100 to 300 mtorr after which the etch rate increases above 300 mtorr pressure. The selectivity between aluminium/2% silicon and the SiO_2 substrate between 100 to 500 mtorr did not appear critical for processing applications. The applied voltages for excitation of a CCl_4 plasma may be higher than that needed for Ar and O_2 plasmas because charge loss by recombination events is probably greater in CCl_4.

One major factor for anisotropic etching is that the effect of the ion bombardment component should be larger than that of the free radical component. However,under the conditions used for etching aluminium/2% silicon films there is a large multiplication process which also produces a large number of free radicals within the dark space near the vicinity of the etched sample. Aluminium is a partially reactive metal (high in the electrochemical series) and will readily react with $Cl^{\cdot}$ free radicals resulting in undercutting of the etched patterns. To reduce this serious undercutting the etching conditions must be shifted away from those favouring multiplication.

Decreasing the pressure of a CCl_4 plasma to 100 mtorr pressure, increases the anisotropic etching component, because the free radical generation will be reduced by moving out of the multiplication region. At the same time there will be a reduction in the aluminium/2% silicon etch rate which can be compensated to some extent by increasing the current to the plasma to 4 Amps rms. However, at 100 mtorr 4 Amps rms the photoresist mask used for patterning the aluminium/2% silicon films deteriorates rapidly, probably because of the higher ion bombardment energy and power used.

Choice of plasma etching is always a compromise as one must try to limit the loss of the masking photoresist layer during etching. If the photoresist layer can be made to withstand harsh bombardment and radical exposure it will make plasma etching processes more flexible.

In addition, there is a need to improve the photoresist resolution by using a thin photoresist layer to obtain submicron structures. This thin patterned layer of photoresist is a problem when plasma etching polysilicon in CCl_4 and Cl_2 plasmas. Methods of reducing the erosion of photoresist masks during plasma etching will be discussed in the final chapter.

CHAPTER NINE

METHODS OF REDUCING THE ETCH RATE OF POSITIVE PHOTORESIST MASKS DURING PLASMA ETCHING

9.1 Introduction

9.1.1 The process for photoresist masks

The fabrication of integrated circuits requires the preparation of defined areas of substrate and surface materials such as polycrystalline silicon, silicon nitride and aluminium. The use of photoresists as masks during chemical and plasma etching is to provide surface arrays on integrated circuits. Such arrays have already been discussed in previous chapters.

The photoresist can be applied to a wafer by spinning a photographically sensitive polymer film (known as 'resist') over the substrate material. This is then heated to 90°C for about 30 minutes to remove the film solvents. The required photoresist pattern is then produced using a mask (usually a selectively etched chromium film on glass) which is placed on the photoresist film and exposed to UV light ($\lambda \simeq$ 400 nm). Photoresist exposed to the light is subsequently dissolved away in a suitable solvent leaving a surface resist mask.

This sequence of pattern transfer to the resist is because the resist used is a positive resist and leaves a resist pattern where the chromium mask prevented the light raaching the resist film. However, resists can be used which crosslink and become insoluble on light exposure, these are termed 'negative resists'.

9.1.2 The criteria for photoresists

A photoresist used in IC fabrication must satisfy the following criteria:-

(i) It must be able to produce patterns at the desired resolution, consistently.

(ii) It must provide an edge profile consistent with the processing requirements.

(iii) It must not introduce unwanted impurities into the etching process.

(iv) It must survive and protect the underlying film during the etching process.

(v) It must be readily removable after the etching process.

Criteria (i) and (ii) can be solved to some extent by using good photoresist polymer systems with high sensitivity (157). Thus with optical projection systems, operating at $\lambda \approx 405$ nm, a resolution of 2.5 μm can be achieved (157). With a step by step projection system a resolution of 1.25 μm has been claimed (157). Finally, using in contact masking (described previously), but with photosensitive resist which responds to lower UV wavelengths, resolution in the submicron region can be obtained (158). Criterion (iii) has been fulfilled because photoresists are now available which do not contain inorganic compounds as fillers or stabilizers.

Criterion (iv) is the most serious limitation on photoresist mask capability for both wet and dry etching. Adhesion problems are often encountered when photoresists are used as a wet etch mask. Two techniques are employed to increase the photoresist adhesion; i) treatment of the substrate surface with an adhesion promoter such as hexamethyldisilazane (HDMS) ii) heating. Plasma etching, however, is not limited by the same adhesion problems as wet etching and (as discussed in previous chapters), planar reactors etch substrate materials with little or no lateral etching component.

The dry etching characteristics of resist and the means of improving their etch resistance are the main subject of this chapter.

Regarding resist removal criterion (v), this can be accomplished using an O_2-plasma, providing there is no inorganic contamination in the resist or any organic substrate exposed to the O_2-plasma.

9.1.3 The plasma process

The sheath and electrode voltages developed in a planar reactor of the Electrotech, 317 type used here, have been discussed previously, in Chapter 8. Photoresist masks used in a planar plasma etcher for etching Al-films in CCl_4 and polysilicon films in CCl_4 and chlorine are subjected to constant bombardment by energetic ions and active components which cause physical sputtering and chemical interactions. Free radicals in the plasma may erode the photoresist mask, particularly at high temperatures. Fig. 9.1 shows a simple diagram of the ion and free radical interactions which can erode a resist layer. The free radical effects in planar reactors are limited provided that the electrodes are cooled. Barrel reactors using free radicals, without wafer cooling and energetic ion bombardment, generally exhibit lower etch rates for photoresist masks than that of planar reactors.

There are three possible ways of improving photoresist mask properties in plasmas:-

i. Free radical quenching

ii. Multilevel masks

iii. Crosslinking the polymer matrix.

It is possible to reduce the free radical etching of photoresist in plasmas by the addition of free radical scavengers to the resist film, such as 2, 5 di-tert butyl p-cresol. Attempts to reduce mask erosion by free radical scavenger additions to a resist can render the polymer mixture unstable without contributing significantly to reducing the resist erosion rate (67).

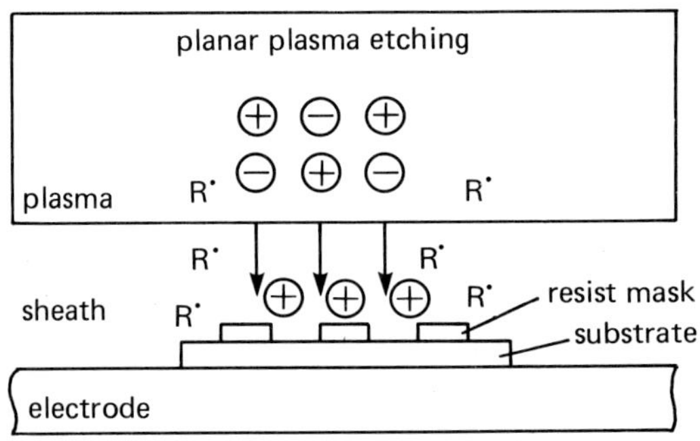

Fig. 9.1

Ion and free radical interactions in a planar reactor.

When a resist mask is shielded from the ion bombardment during plasma etching but allowed free radical access, it had a much lower etch rate than those areas exposed to ion bombardment. This result indicates that free radical etching alone is a very small component of the etching process in a planar system (67).

Multilevel resist (MLR) methods have been used with some success (160). MLR methods can give higher height to width aspect ratios compared with a one layer resist pattern and are classified by the number of layers used, e.g. (i) the bi-level i.e. double layer (ii) tri-level i.e. triple layer. Both methods start by using a thick resist or polymer layer which has the advantage of planarizing any high steps over parts of the wafer topography.

This thick resist layer can be patterned by coating and printing a thinner resist layer over it. If the thick resist layer is a PMMA type polymer and the thin resist patterned layer is a Novalac type polymer then on flood exposure to deep UV radiation (λ = 254 nm), the Novalac resist will mask the radiation. The PMMA that is exposed to this radiation can be dissolved using iso-propyl alcohol and iso-butyl methyl ketone leaving a pattern consisting of two levels of resist. In method (ii) (shown in Fig. 9.10b) the first layer is usually a thick Novalac resist onto which a thin metal or silica layer is deposited. A thin resist layer is patterned over the metal or silica layer and the exposed areas of metal or silica are etched away leaving the thick resist exposed in those areas. The metal (usually aluminium) or silica are good masks for reactive ion etching using an O_2-plasma to remove the thick resist layer. This method can produce fine lines in thick resist layers over large steps (35).

MLR methods have a major disadvantage of requiring additional process steps and leads to associated problems. In the bi-level method the two different resist layers have been known to separate (67). With the tri-level method, debris sputtered from the electrodes during reactive ion etching causes process problems. During the reactive ion etching (RIE) in an O_2-plasma to remove the thick resist layer, the substrate becomes exposed when the resist is etched through. If the substrate is aluminium then it reacts with the high energy oxygen ions, this causes longer induction periods during plasma etching of the exposed aluminium substrates (67).

9.1.4 Cross-linking and scope of work

In this chapter research work will be described which was undertaken to improve the etch resistance of resists using cross-linking of a printed photoresist mask as a pre-treatment before plasma etching. Initially cross-linking the photoresist mask layer,and thus forming a higher molecular weight polymer matrix was expected to improve the photoresist mask life in the plasma. Mask life should be prolonged by cross-linking, as the number of bonds to be broken (mainly by ion impact), before the photoresist film can be made volatile, is increased. In addition,cross-linking can sometimes produce larger bond energies in the polymer, making it more difficult for bonds to be ruptured. Cross-linking will also decrease the photoresist free radical etch rate because a larger number of free radicals arriving at the resist surface can be tolerated before complete volatilisation. The extent and type of cross-linking will depend on the method used to achieve it, and on the type of polymer. The change in the erosion rate of a resist after cross-linking will also depend on the subsequent ion bombardment conditions, because an ion may rupture more than one bond in the polymer matrix depending on its energy and type. Once a bond is ruptured, free radicals may react rapidly and erode the polymer film.

Cross-linking methods tried for reducing the etch rate of the resist mask during plasma etching were:

i. Heat pre-treatment
ii. Ultraviolet exposure
iii. Plasma pre-treatment in inert gases
iv. Chemical pre-treatment
v. Ion implantation using Ar^+ impact
vi. Electron beam irradiation

The effectiveness of these pre-treatments and the resist loss arising from the pre-treatments are discussed. For VSLI requirements the choice of photoresists is limited to those capable of fine line resolution. For optical lithography, positive photoresist systems generally have higher resolving power than negative resists of the same thickness. The resists investigated here were mainly HPR 206 and 204 and others such as Macdermid PR74. Some resists which respond to deep UV i.e. short UV wavelength such as HPR WX159 and PMMA were also used for comparison studies. Infra-red analysis on the resist films was used to monitor changes during pre-treatment and etching.

9.2 Background to Photoresist Chemistry

Positive photoresists consist of a phenolic type resin and a diazoquinone sensitizer which are both dissolved in a solvent. Small amounts of filler (solids) and antioxidents are added to the photoresist system to improve the storage life and coating properties.

On exposing the photoresist film to radiation the aromatic diazoquinone sensitizer is photolytically induced to undergo intra-molecular decomposition and rearrangement (161). Referring to Fig. 9.2, the steps of the reaction are:-

a. The light sensitive diazo group absorbs radiation forming a 'carbene' with loss of nitrogen.

b. The carbene is unstable and rapidly undergoes a ring contraction reaction to form a 5 membered ring known as a 'ketene'.

c. The ketene reacts with absorbed water in the photoresist film to form a 'carboxylic acid'.

d. The carboxylic acid formed allows the dissolution of the insoluble resin system in an alkaline developing solution.

Photoresist films contain between 0.1%-1.0% of absorbed water and any unreacted ketene will react in the developing alkaline solution or the Novalac polymer to form esters (161). The sensitizer (diazoquinone) in positive photoresists is not only photolytically unstable but is also thermally unstable. The moderate bake temperature at which the solvents in the photoresist film are removed must be below $\sim 100^oC$, because above this temperature the rate of degradation of the sensitizer becomes significant. The moderate bake temperature is known as a 'soft bake' during resist processing.

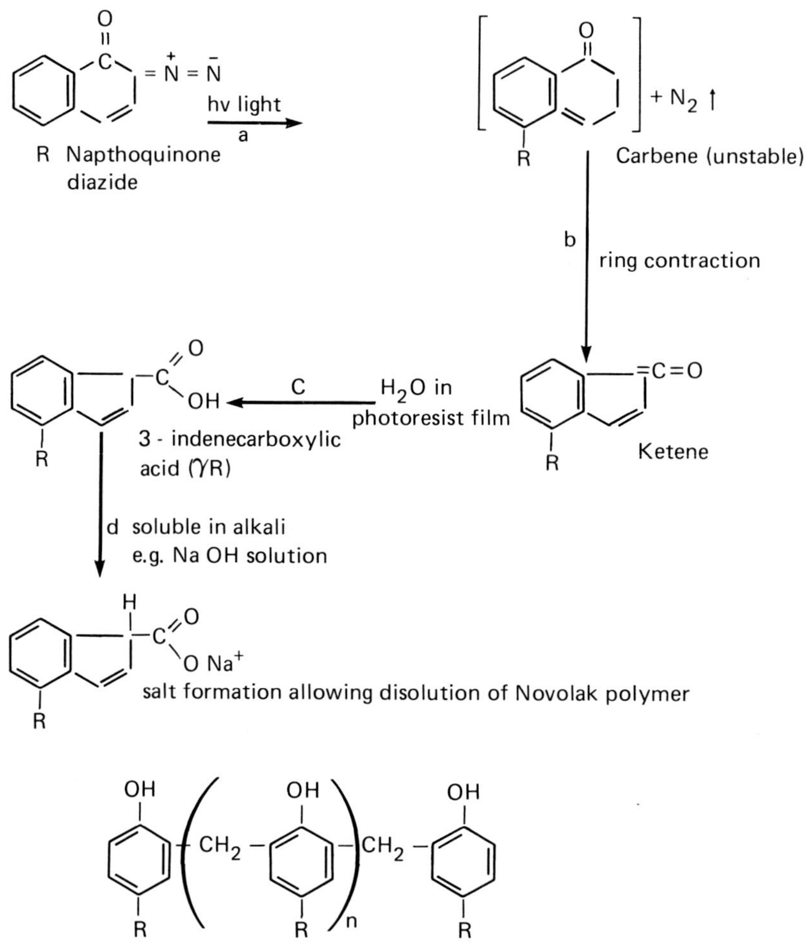

Novolak polymer
n is between 0 – 13 but on average is 5
R may also be substituted in the (m) or O – positions

Fig. 9.2
Reaction steps of a positive photoresist.

A photoresist film contains ∿ 20% sensitizer, some of which is lost during the soft bake and this can reduce the sensitivity of the photoresist film. However not all the sensitizer need be photolytically decomposed to allow dissolution of the resist system via the alkali soluble carboxylic acid process.

Negative photoresists work by a different mechanism to that of positive photoresists. In this case differential solubility is achieved via a photolytically induced inter-molecular cross-linking reaction. Negative photoresists have a larger polymer matrix which improves the resistance to thermal flow (deformation) and chemical attack. Negative photoresists however tend to swell in organic solvents used for the development process and must be 'deswelled' by the appropriate rinses to prevent loss of resolution.

Positive photoresists with a lower average molecular weight and narrow molecular weight distribution have better resolution, steeper resist profile, are less viscous and in some cases display better adhesion. By the correct control of the molecular weight distribution resists have the capability of forming 'striation free films'. However, as positive photoresists have a lower average molecular weight compared with developed negative resists they are less resistant to chemical attack and thermal flow. Moreover, the molecular weight control needed for positive photoresists makes their manufacture expensive compared to that of negative photoresists.

The sensitizers in positive photoresist respond to a broader spectrum of radiation than the bis-azide sensitizer used in conventional negative photoresists. This is shown in figs. 9.3 and 9.4 which compare the optical absorption of positive and negative photoresists respectively. The positive photoresist has an extended absorption above 400 nm. This is an advantage in projection printing, because the

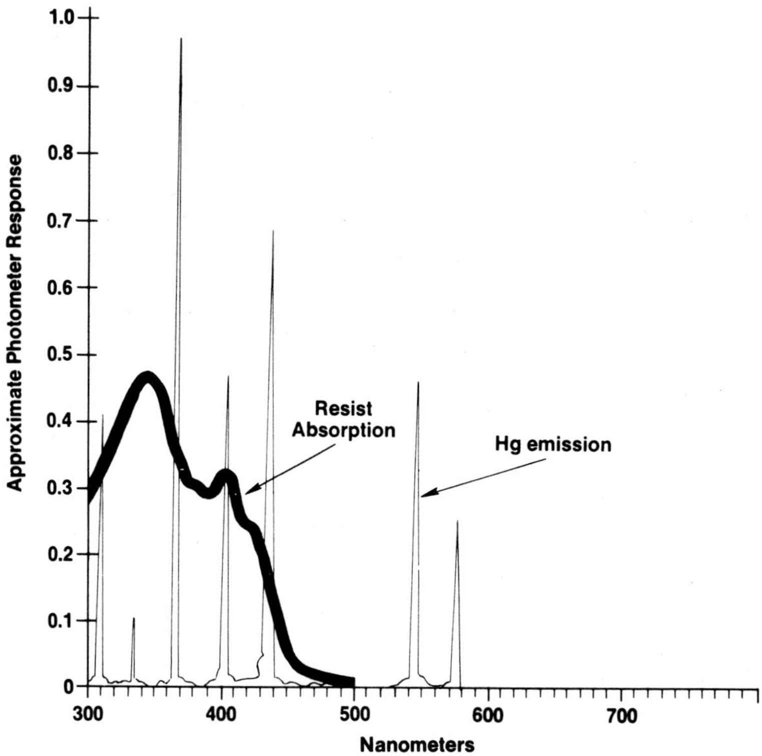

Fig 9.3
Absorption Spectrum of a Positive resist (from ref 159)
Before exposure

apparatus uses the 405 and 436 nm emission lines from high pressure Hg lamps (in most aligners) for imaging rather than the 365 nm radiation required by the bis-azide sensitizer. The ability to use the 405 and 436 nm emission lines can reduce the exposure time and reduce the standing wave patterns which are due to the light being too coherent. A summary of the properties of positive and negative photoresists is shown in table 9.1.

TABLE 9.1

SUMMARY OF THE PROPERTIES OF POSITIVE AND NEGATIVE PHOTORESISTS

	Positive Resists	Negative Resists
Molecular weight average in film	1,000	400,000
Polymer System	Novolaks and Resale + Others	Cyclized poly-cosisoprene and polyringed cinnamate
Type of sensitizer	diazo ketone responds between 340-460 nm ~ 20%	Aromatic Diazide responds around 365 nm ~ 2%
Solvents Normally used in resist	Xylene and 2. Ethoxy ethyl acetate	Xylene, Cyclo-hexanone and Butyl acetate.
Chemical Resistance of film	Poor:- easily oxidized and soluble in strong alkaline/ acid solutions	Good
Development process and adhesion	Solubility differential achieved in alkaline solutions, can have adhesion problems because molecular weight too high to achieve low film stress.	Solvent deswell process extraction. The low uncrosslinked polymer adhesion is good due to low molecular weight before exposure.
Problems and Resolution	Better resolution can normally be achieved in positive resists than negative resists. Exposure for positive resists is usually slower and suffers less from humidity problems than negative resists. Water is required in positive resists in order to work see fig 9.2	In negative resists cross-linking decreases at the edge of the image. Spreading of the image can occur if overexposed. Negative resists suffer from standing waves during printing because of the coherence of the light source and the sensitizer being tuned to one particular wavelength from the source. Negative resists suffer more from humidity problems because water can render the small amount of aromatic bisazide (~2%) inactive.

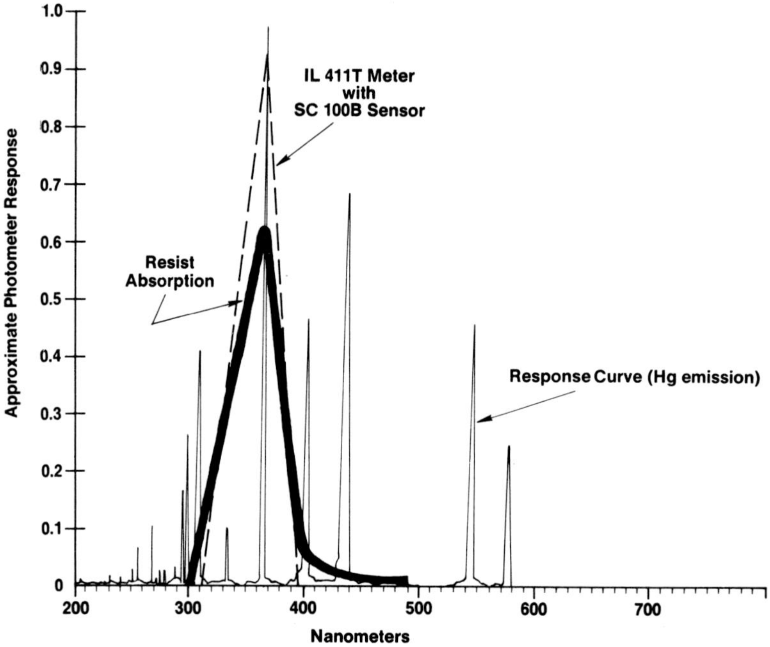

Fig 9.4

Absorption Spectrum of a negative resist (from ref 159)
Before exposure

9.3 Experimental Work

9.3.1 Wafer construction

a. Aluminium coated Si-wafers were prepared from 3 in. dia. bulk silicon (1-17 Ω cm) onto which an 0.5 μm thick layer of thermal oxide was grown before aluminium deposition. The oxide covered wafers were immersed in a 5% HF solution for 15 seconds before depositing a 1 μm thick Al-film containing 2% Si using a magnetron sputtering system (CVC model).

b. Si-wafers coated with polysilicon were prepared from 3 in.dia. bulk silicon of very high resistivity (1000 Ωcm) used to reduce absorption in the infra-red region for spectrometry work. A thin silicon dioxide layer (∿ 350Å) was grown onto the bulk silicon prior to polysilicon deposition to prevent the exposed bulk silicon from etching during the plasma etching of the polysilicon. The oxide layer was necessarily thin to prevent appreciable absorption by the oxide in the infra-red region. Deposition of polysilicon (0.5 or 1 μm thick) was carried out using a low pressure chemical vapour deposition system.

The wafers were then coated with photoresist usually HPR 206 (∿ 2 μm thick) and baked at 84°C for 5 min and then contact printed using a mask type 407 described next.

9.3.2 Mask 407

Mask 407 was a high resolution test mask specially designed for this photoresist investigation. The mask contained various patterned shapes and sizes (down to 0.8 μm minimum feature size) to test the photolithography and etching processes. The mask 407 was arranged in chip die formation and had an exposed substrate area between each chip pattern row, thus making the mask 407 more representative of typical masks used in IC fabrication. Using mask 407 it was easier monitor visually the end points during plasma etching than the M.I.T.* test pattern which has only a small substrate area exposed to the plasma.

In the centre of the mask was an area of 1.2 × 2 cm. which left a blank area of photoresist film and when using the low doped/high resistivity silicon, this was the region used for infra-red analysis.

*M.I.T. Micro Image Technology, U.S.A. High resolution test mask.

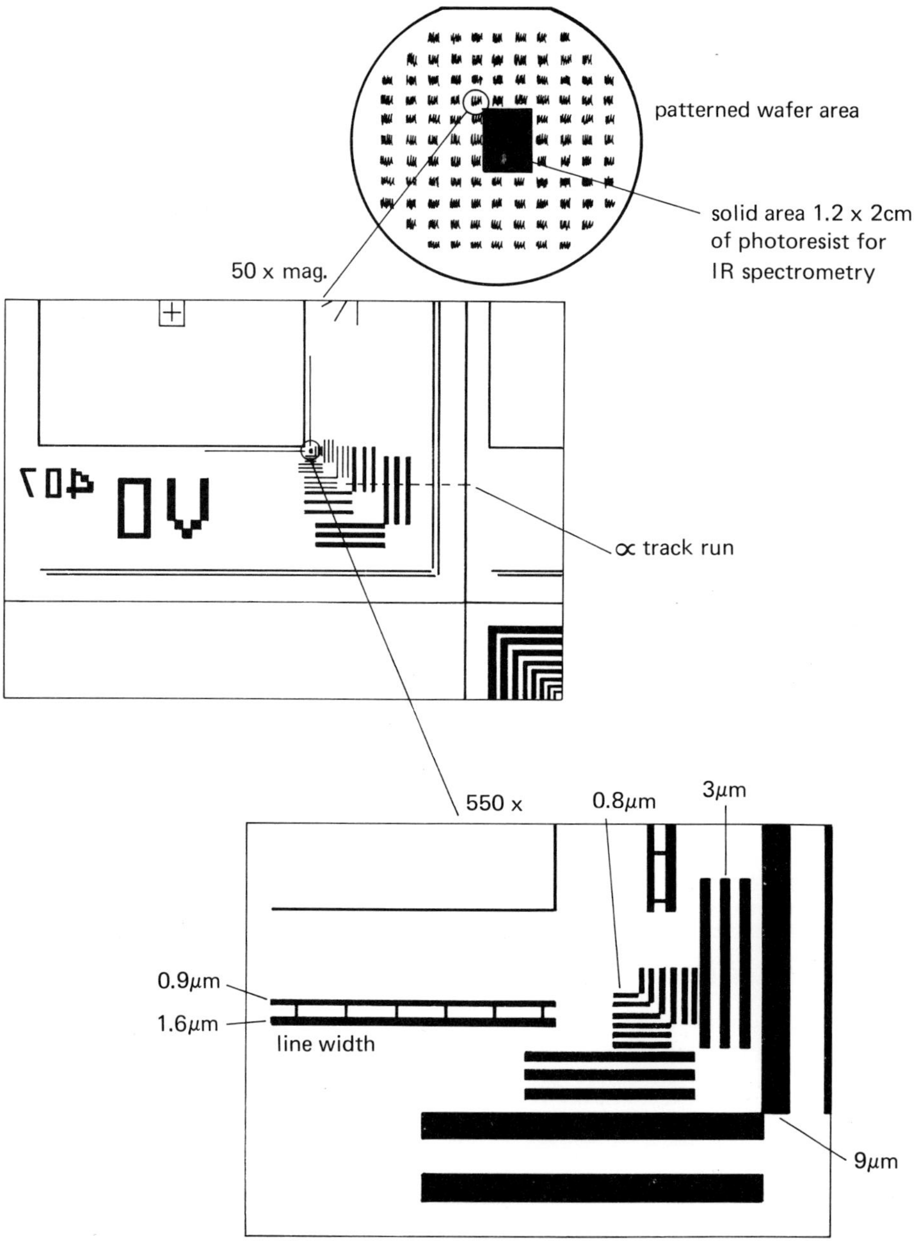

Fig. 9.5
Mask 407 patterns and dimensions.

The infra-red analysis on the resist film was used to monitor changes during pre-treatment and etching. The infra-red apparatus used was the double beam Perkin Elmer spectrometer type 580B with data station. Fig. 9.5 shows the shape of the patterns on mask 407 and the size of some of the patterned lines.

9.3.3 Measurement of the resist loss

The resist film thickness was measured at specific locations on the wafers before and after cross-linking pre-treatment and after plasma etching. The apparatus used for thickness measurement was a Tenco α-step. This apparatus is similar to a 'Taly Surf' i.e. thickness measurements are obtained by monitoring the up and down movement of a stylus as it passes across the wafer surface. However, the α-step has ten times more sensitivity and a low load on the stylus to prevent scratching off the resist. The measurements were made across particular regions of the patterned lines as shown in fig. 9.5 by the dotted line. Thickness measurements were taken at the specified places as follows:-

i. Before pre-treatment

ii. After photoresist pre-treatment

iii. After plasma etching

iv. After resist removal to measure the etch step made by the plasma etching

From these results resist loss during pre-treatment and etching was found. During plasma etching processing large variations in thickness measurement from run to run can arise from loading effects, temperature changes and residual gas contamination. In addition the resist etch rate in a plasma is not constant because of the resist outgassing and temperature rise. To reduce the run to run changes in etching conditions, the photoresist films which had been exposed to various pre-treatments were etched together with up to five wafers per run. By having the various pre-treatments etched in one run a better comparison can be made. Although this reduces the run to run variation there are still etching variations across the reactor and across the wafer. The Cl_2-plasma process gives poor wafer uniformity resulting in large deviations in the resist measurement after plasma etching.

The etching process was applied under the same conditions as used on the device wafers. An Electrotech 317 planar reactor (15) was used with a CCl_4 plasma at 2 Amps (r.m.s.) current and 300 mtorr pressure to etch 1 μm thick Al-films with 50% overetch. The 50% overetch means: after etching through the Al-film (monitored visually) in say 10 min, a further 5 min of etching time was given to ensure complete removal. For the 0.5 and 1 μm polysilicon layers a Cl_2 plasma was used under the same current and pressure conditions but with a 25% overetch.

9.4 Discussion of Pre-Treatments

After several runs the resist etch rate for each pre-treatment of HPR 206 is summarised in Fig. 9.6 and Fig. 9.7 for aluminium and polysilicon etching respectively. The lower line in Fig. 9.6 indicates the amount of resist loss that occurs during the various pre-treatments.

9.4.1 Heat pre-treatment

Photoresists have a glass transition temperature at ∿ 120°C and above this temperature will begin to flow causing bevelling, loss of line width control and resist thinning. This is demonstrated by the SEM photos in figs. 9.8 and 9.9, which show resist patterned lines 3 μm wide that have been heat treated for 15 min at 120 and 150°C respectively.

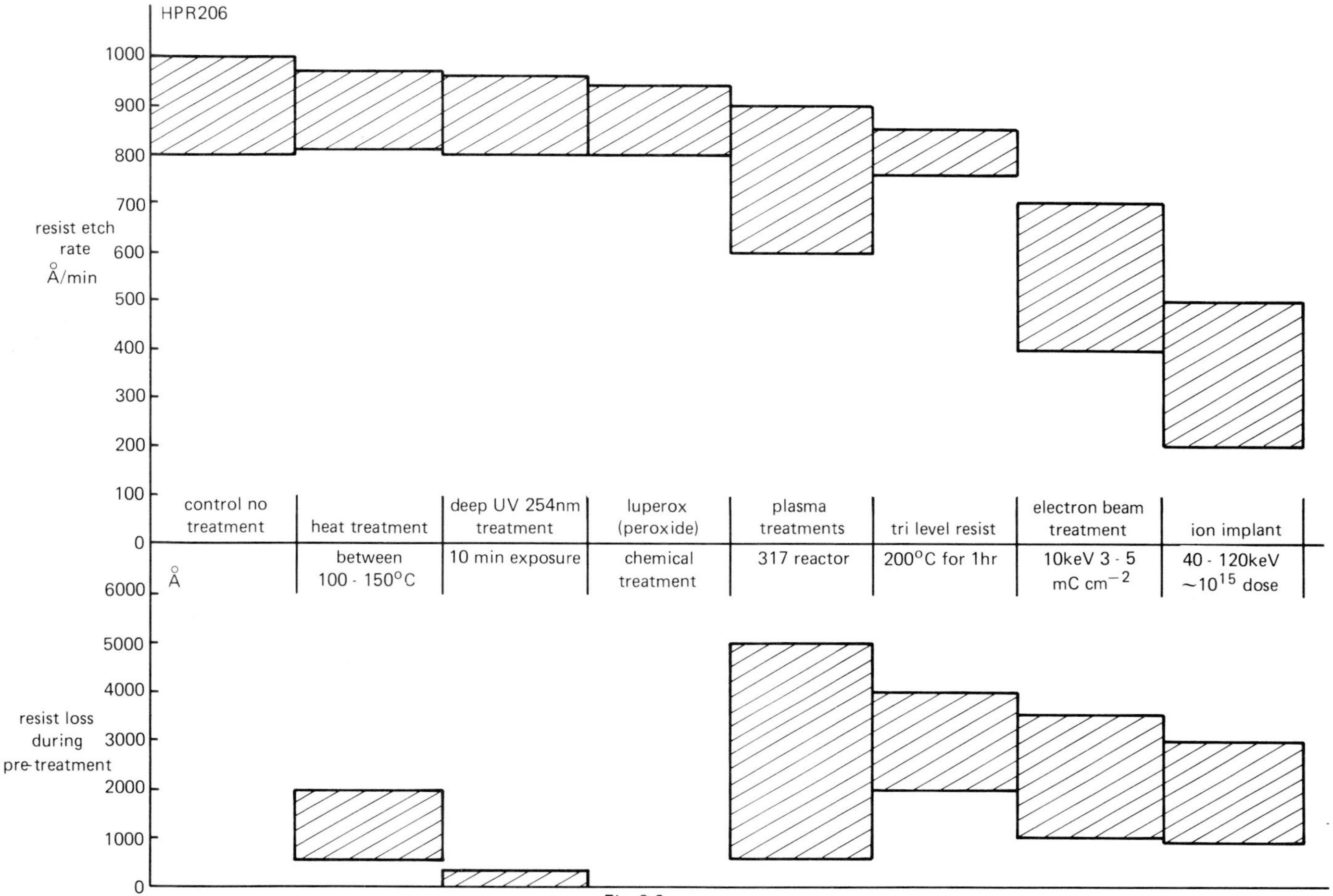

Fig. 9.6
Summary of resist etch rate for various pre-treatments during the etching of aluminium in a $CC1_4$ plasma at 300 mtorr, 2 amps.

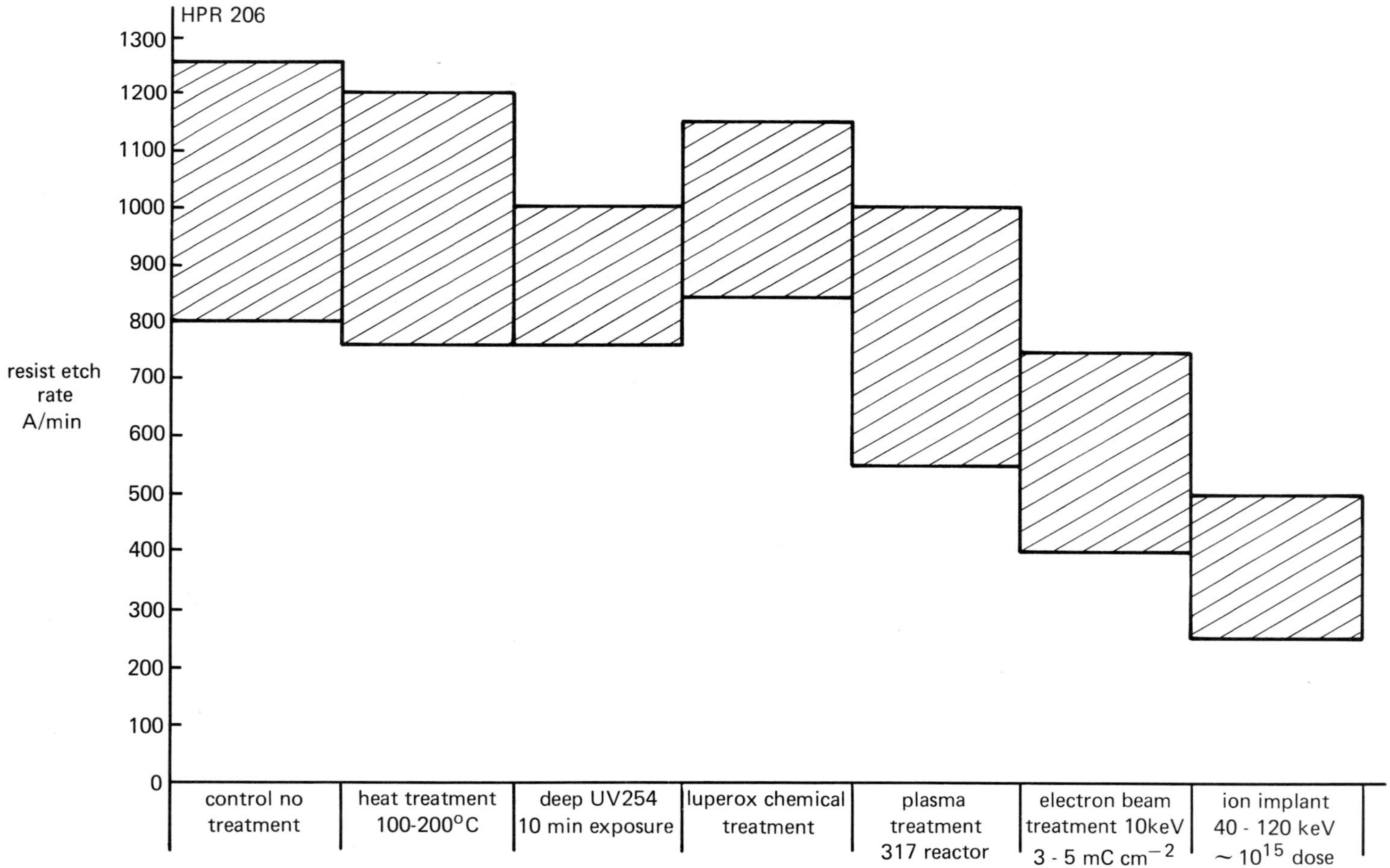

Fig. 9.7

Summary of resist etch rate for various pre-treatments during the etching of polysilicon in a Cl_2 plasma at 300 mtorr, 2 amps.

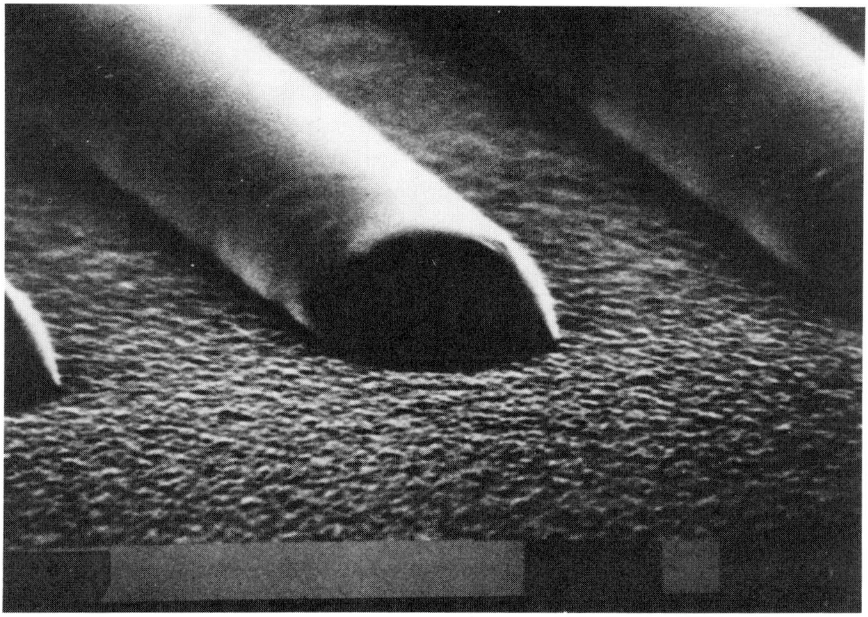

Fig. 9.8. SEM photograph of a 2 μm thick HPR206, with a 3 μm wide line heated to 120°C for 15 min.

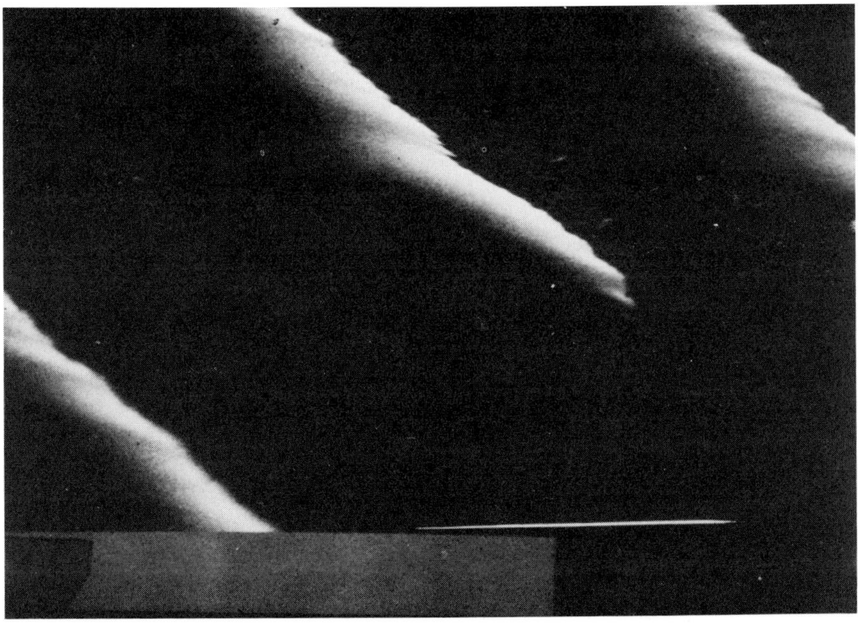

Fig 9.9. SEM photograph of a 2 μm thick HPR206, with a 3 μm wide line heated to 150°C for 15 min.

Heat pre-treatment (< 200°C) produces only limited cross-linking whilst the flow induced can cause considerable changes in the resist pattern profile. These changes in the resist pattern profile make step height measurements for resist thickness difficult. This is shown by the α-step trace resist thickness measurement of a sample before and after heat pre-treatment, as illustrated in Fig. 9.10a. For the heat treatment study, measurements of the resist pattern thickness is best prepared by steps (1) to (7). (1) Coat resist over wafer surface, (2) heat treatment for ½ hour (no flow will be observed since there is no defined pattern so far), (3) evaporate 1000A of Al onto the resist layer, (4) coat and pattern a resist layer (Mask 407 in contact) on top of the thin Al layer, (5) wet etch the exposed Al layer, (6) RIE in an O_2 plasma the exposed underlying heat treated resist layer and (7) wet etch away the remaining Al mask. The foregoing technique, known as the tri-level process, is shown in its stages in Fig. 9.10b. The samples can now be used in plasma etching studies and the results of the heat pre-treatment using this method are shown in Fig. 9.11. There is a significant reduction in the resist etch rate in a Cl_2-plasma after heat pre-treatment at about 250°C. However, the resist becomes difficult to remove after the plasma etching process if the resist is heat treated above 200°C.

Differential thermal anaylsis (DTA), thermal gravimetric analysis (TGA) and infra-red spectrometry of photoresist films has been previously reported by the author (162). Fig. 9.12 is a TGA of a HPR 206 film and shows the weight loss of the resist film during heating in air, which indicates similar volume losses to to those shown in Fig. 9.11. The DTA data of HPR 206 resist is shown in Fig. 9.13. This reveals two exothermic reactions; a large exothermic point at 140°C which is due to the sensitizer decomposition and a small exothermic reaction at 250°C

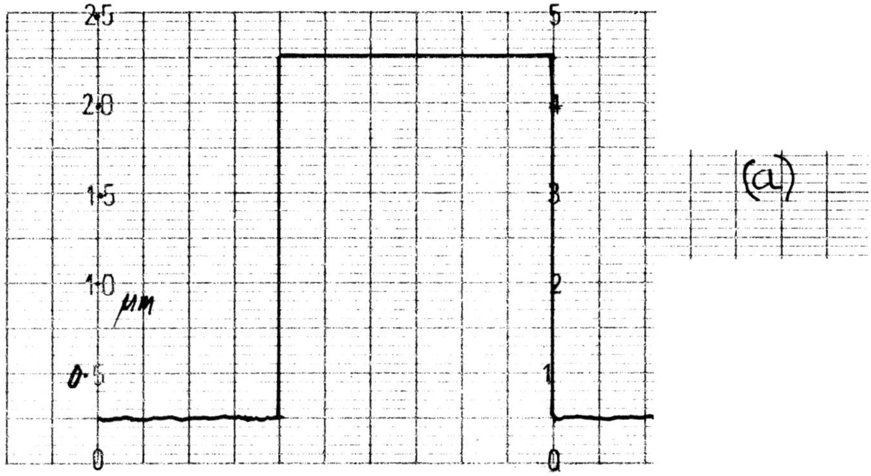

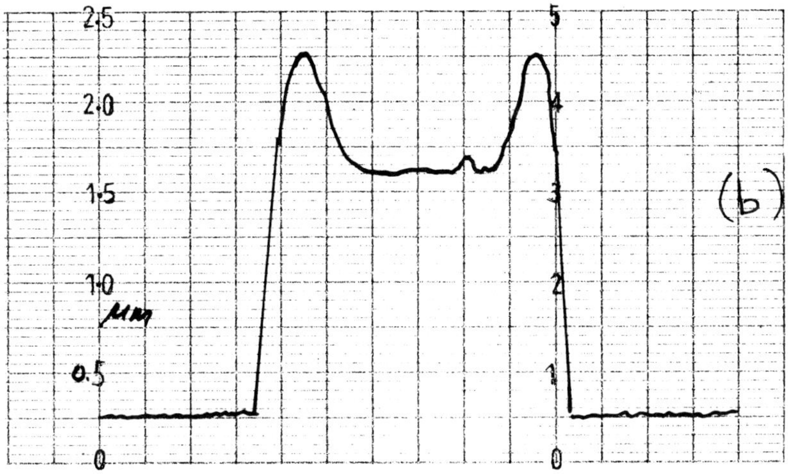

Fig 9.10a

Distortion in the resist profile using an α-step plot, (a) before heating (b) after heating to 150° C for 10 min.

which could arise from cross-linking in the Novalac polymer system. The latter could explain the reduced etch rate observed, and shown in Fig. 9.11 Infra-red spectrometry revealed a small increase in C-O absorption after heat pre-treatment of the resist film. The rise in C-O absorption in the IR-spectra of a resist film after heat treatment is shown in Fig. 9.14. The C-O linkages may be due to cross-linking of the Novalac polymer

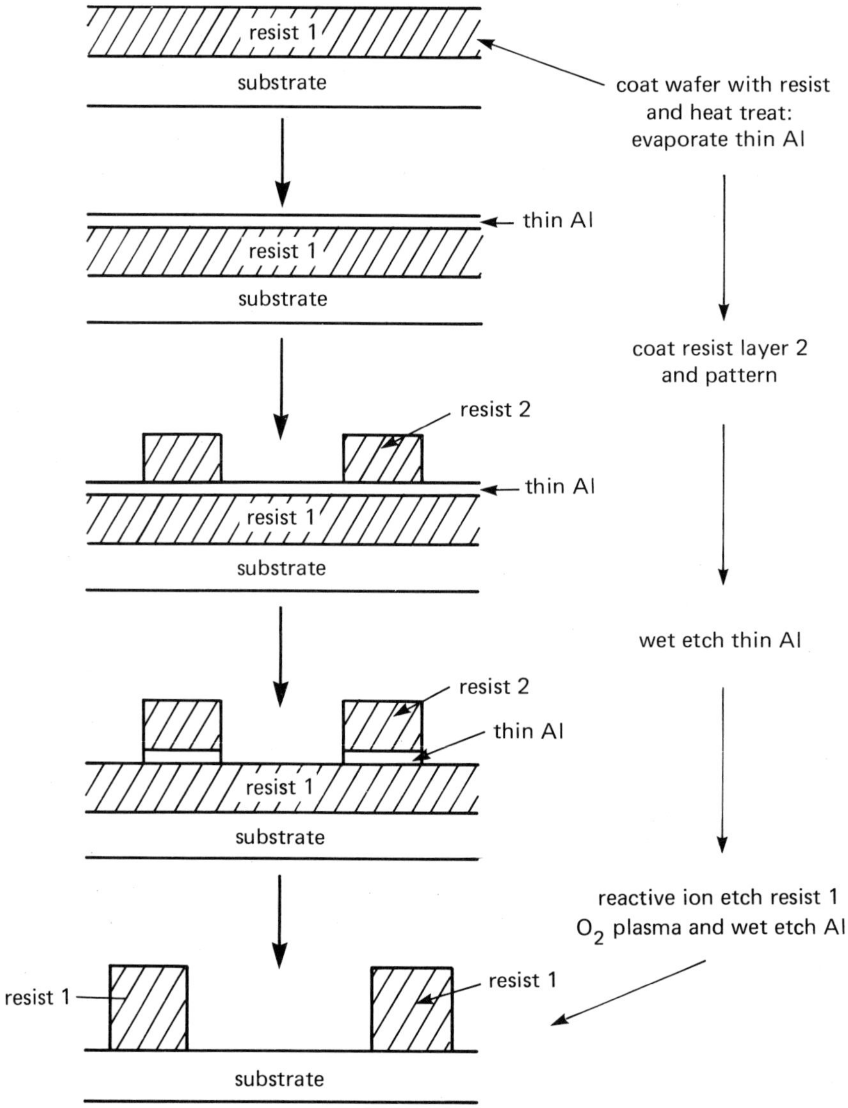

Fig. 9.10b
The trilevel process

system as shown in Fig. 9.15. Although heat pre-treatment below 200°C does not improve significantly the etch resistance of photoresist, it can reduce pin hole formation and solvent loss and reduce N_2 evolution from the resist layer during etching (162).

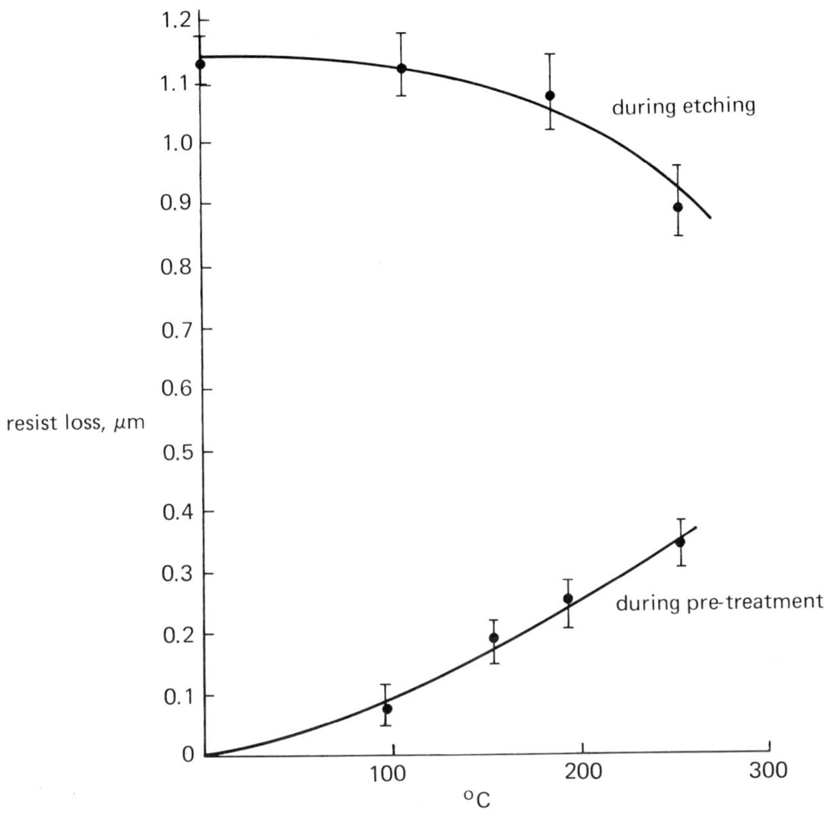

Fig. 9.11

Top:- Plot of resist thickness loss during polysilicon etching using a Cl_2 plasma vs resist bake pre-treatment temperature.
Bottom:- Plot of resist thickness loss during pre-treatment vs pre-treatment resist bake temperature.

9.4.2 Plasma pre-treatment

Resist coated wafers were patterned by contact exposure using the mask 407. The wafers, after patterning, were given a plasma pre-treatment in a planar reactor (Electrotech 317) using an Ar or He plasma at 2 Amps r.m.s. current and at various pressures. Some wafers were rested on the ground electrode and others used were electrically floated by resting the wafer on an insulator placed in the chamber.

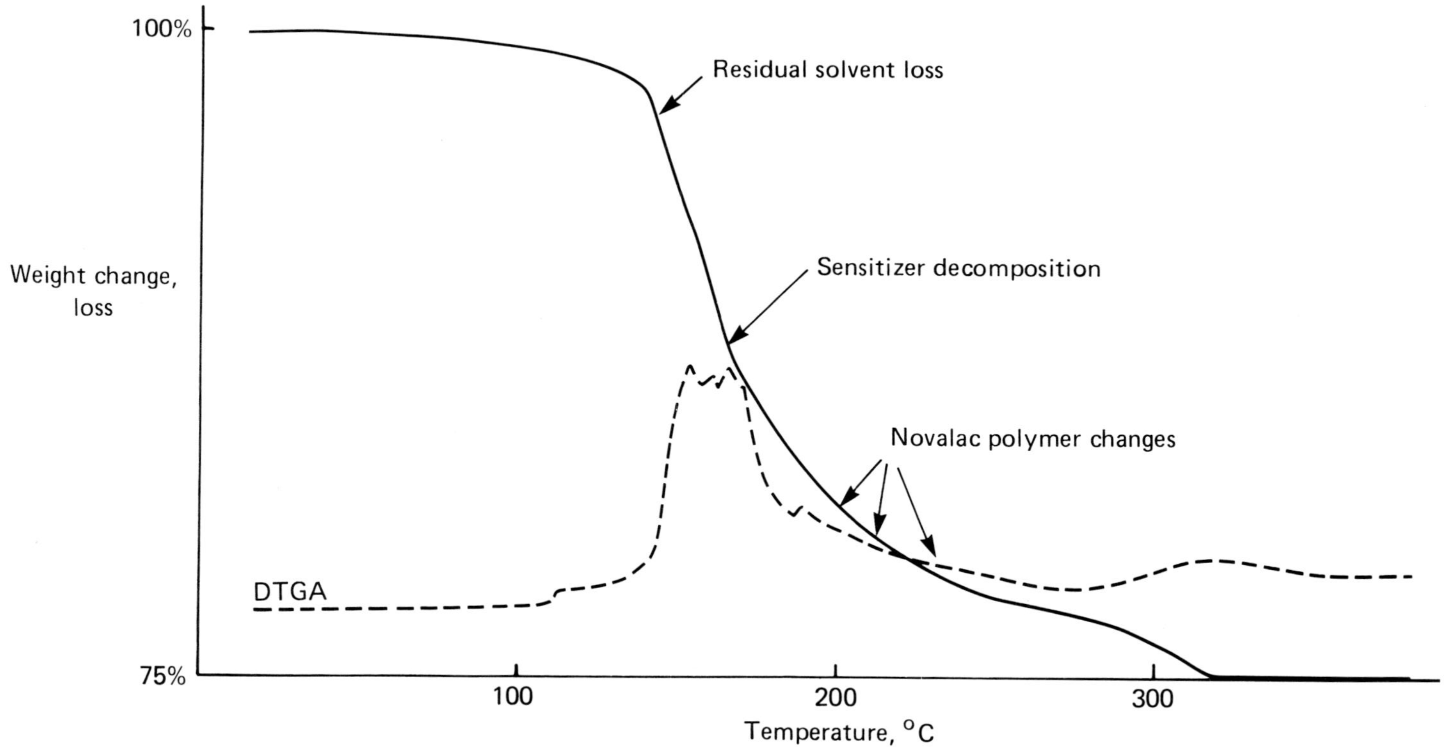

Fig. 9. 12

Dupont Thermal Gravimetic Analysis of a HPR 206 film ATM, air 160 cc/min.

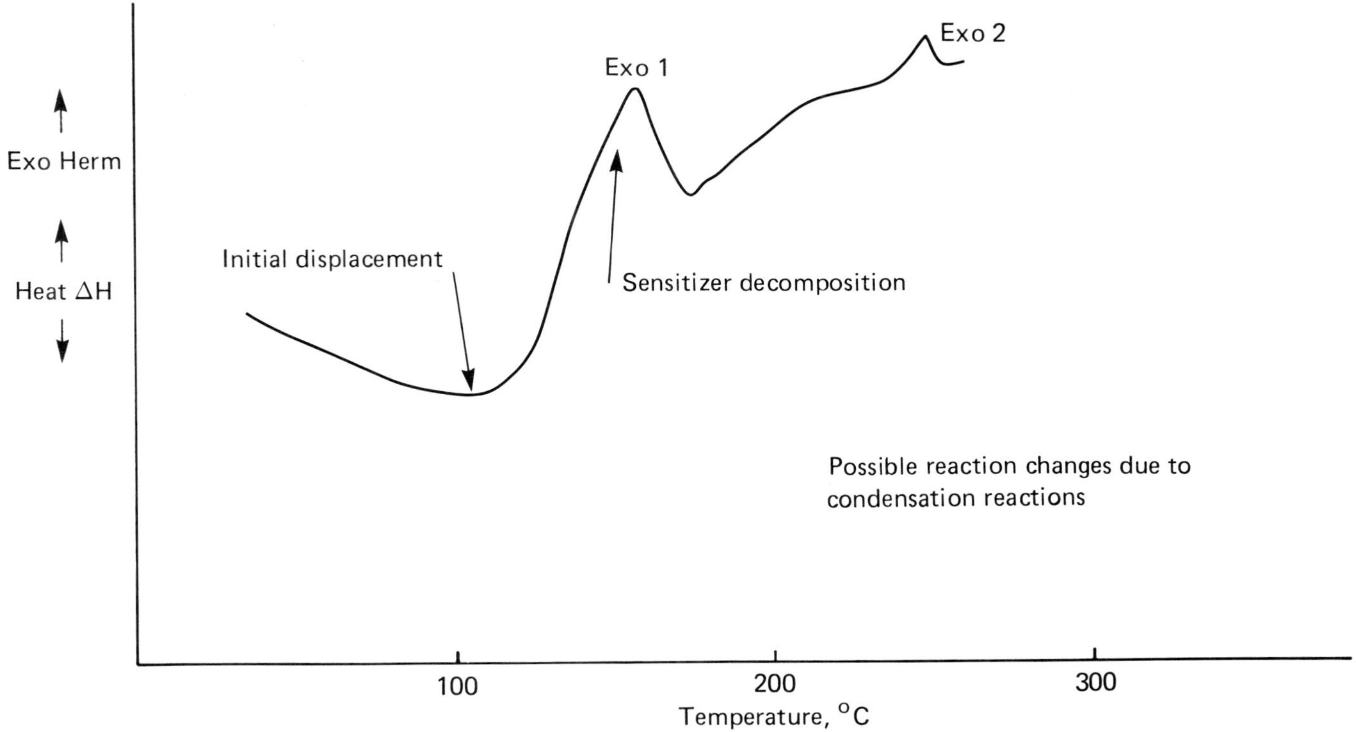

Fig. 9.13

Dupont Differential Thermal Analysis of a HPR 206 solution ATM, air flow rate 160 cc/min.

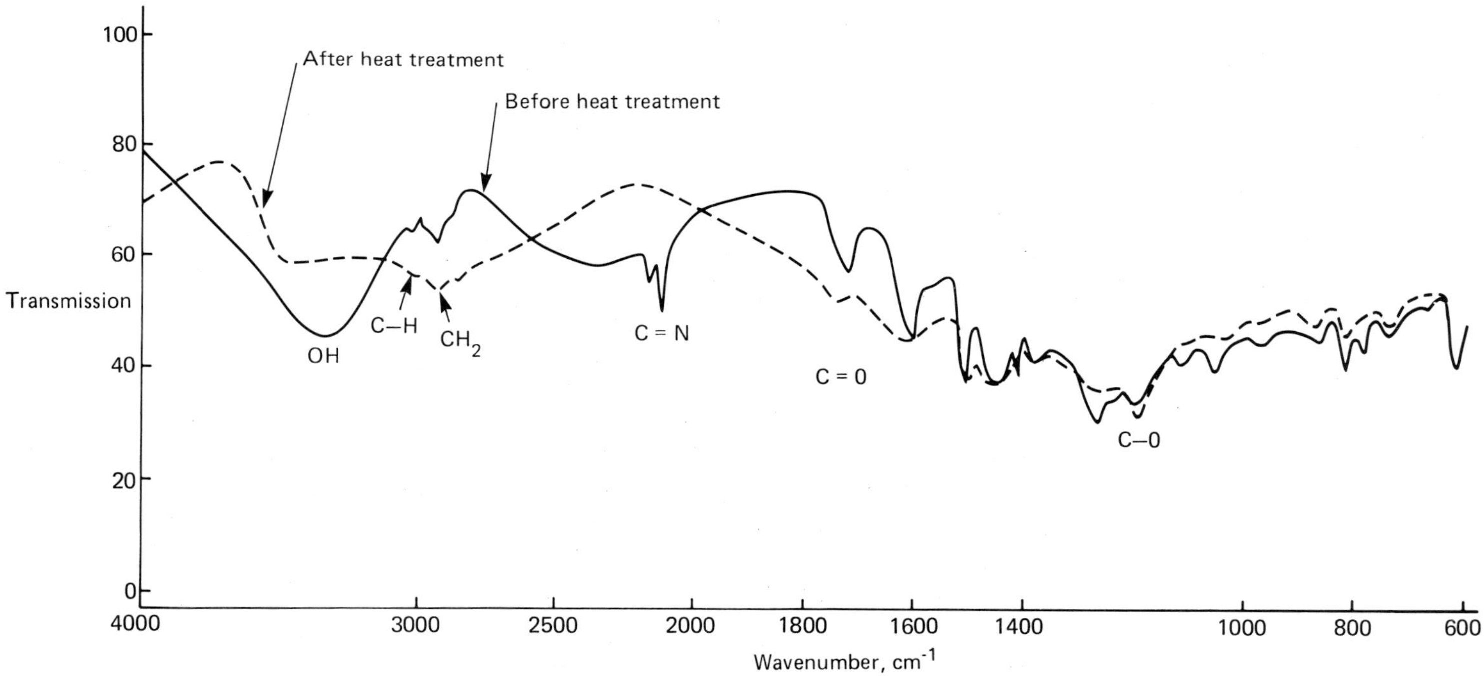

Fig. 9. 14

Comparison between untreated HPR 206 resist 2μm thick and then heat treated at 160°C for ½hr.

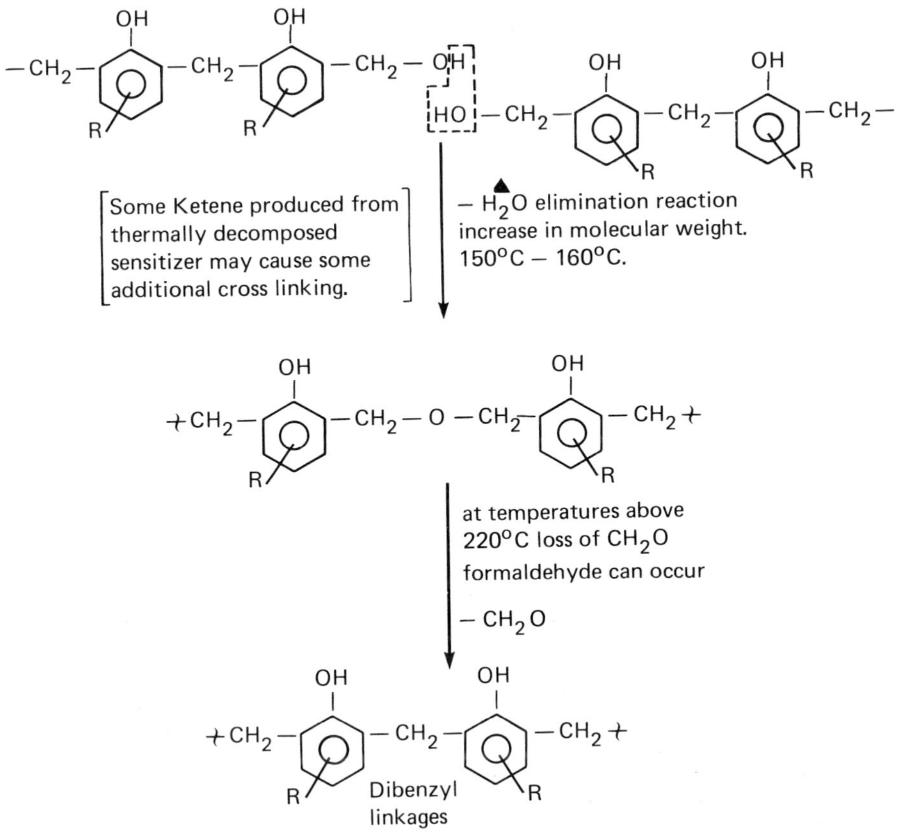

Fig. 9. 15

Possible thermal cross-linking mechanisms of Novolak polymers.

A low pressure of 100 mtorr, using Ar or He plasmas with wafers resting on the ground electrode (for between 10 to 20 minutes), gave the best results for lowering, (albeit small) the resist etch rate during plasma etching. However at these conditions the loss of the resist thickness during pre-treatment to promote resist hardening was often considerable. When resist wafers are resting on the ground electrode, the large ion bombardment energies produced in Ar or He plasmas at 100 mtorr can cause sputtering of the resist. This, coupled with small air leaks in the He and Ar line or chamber, and possible desorption, i.e. memory effects from previous etching runs, could produce the observed large resist loss and run to run variation during plasma pre-treatment. The chamber, however, was always pre-conditioned by running an Ar or He plasma before pre-treatment. By electrically floating the resist wafer in the plasma the ion bombardment energy is reduced, resulting in less resist loss during plasma pre-treatment. However this floating method did not significantly reduce the resist erosion rate during plasma etching.

Although all plasma pre-treatments did not improve the resist erosion rate significantly, it can stop the resist from flowing at high temperatures, i.e. image stabilization, as shown in Fig. 9.16. Fig.9.16 is a photograph of a patterned area after plasma treatment, which has been subsequently heated to 180°C for 1 hour. The wafers were pre-treated on the ground electrode, but similar results are obtained from floating wafers in the plasma.

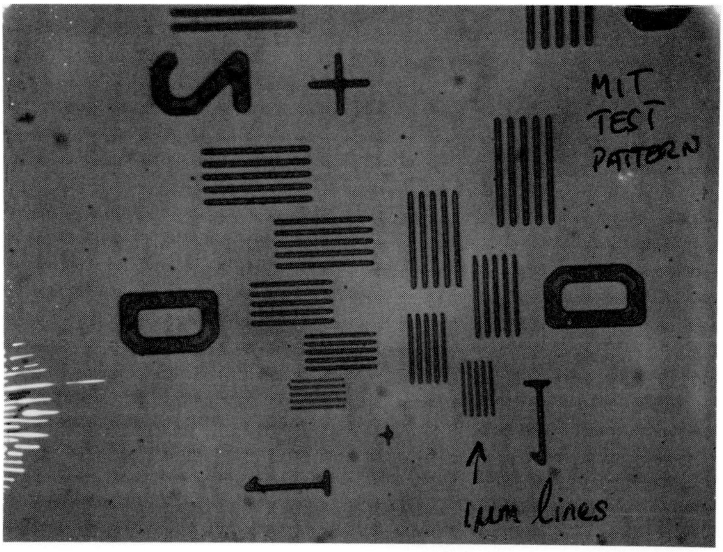

Fig 9.16

Photograph of a patterned area which has been plasma pre-treated in a planar reactor and then heated to 180° for 1 hour.

The surface temperature from bombardment of a resist layer during plasma pre-treatment, floating or even on a water cooled electrode may reach > 120°C (67). This did cause wrinkling in the resist film probably because of the difference in thermal properties of the cross-linked surface 'skin' and the underlying resist layer. The skin produces more wrinkling on rapid heating. This effect of wrinkling can be reduced with slow thermal rises. The skin formed over the resist layer during plasma pre-treatment is probably caused by ion bombardment inducing cross-linking at the resist surface. Because some of the wafers were floating during plasma pre-treatment, the ion bombardment energy will be lower and may not contribute to skin formation. However the UV produced in the plasma may cause some cross-linking i.e. skin formation at the resist surface. If, after plasma pre-treatment, the resist wafer is placed in an O_2-plasma using an Electrotech 317, at 100 mtorr pressure, 2 Amps r.m.s. current for 30 seconds, the skin is removed and the resist flows on heating above 130°C using a hot plate (162). Thus the cross-linked skin over the resist surface prevents the resist from flowing at high temperatures.

Pre-treatment in a planar reactor is more effective than pre-treatments in a barrel reactor for giving small reductions in the plasma resist etch rate. This is probably because the additional ion bombardment in a planar reactor helps to cross-link the resist surface. Thus ion bombardment pre-treatment is the main source of reducing the subsequent etch rate of the resist and of stopping flow at high temperature.

Experiments were made using a barrel reactor (International Plasma Corporation) with shield. The resist wafers were exposed to CF_4, N_2 and He plasmas each at 700 mtorr pressure and 400 watts r.f. power input for 15 min. This pre-treatment does not reduce the resist etch rate in subsequent plasma etching and does not reduce the resist flow on

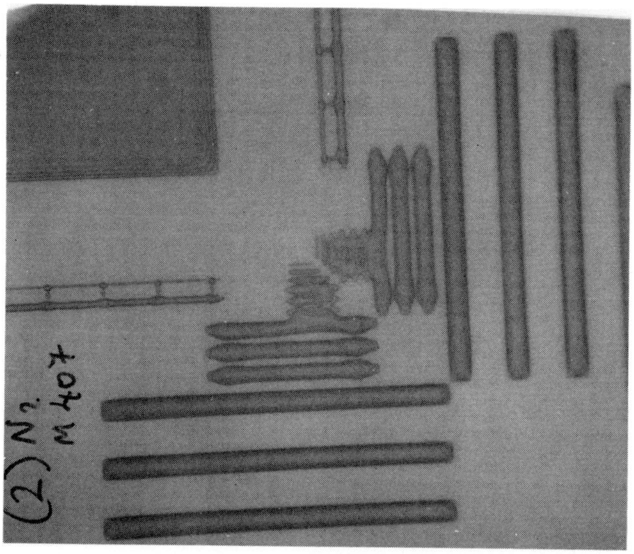
(2) N2
M407

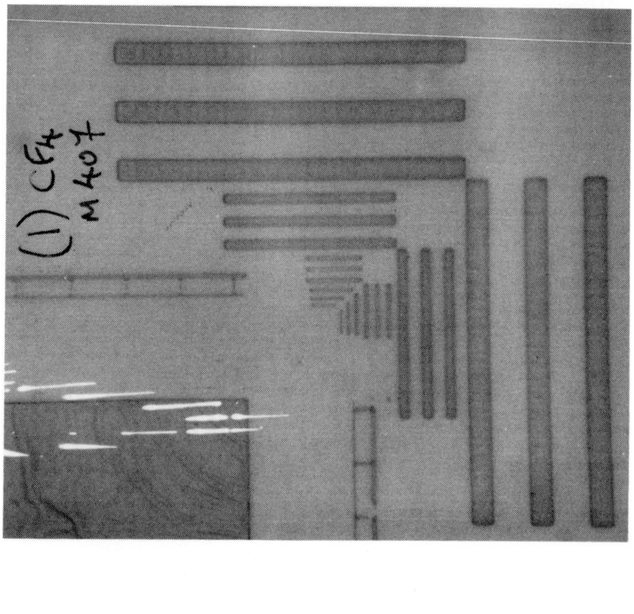
(1) CF4
M407

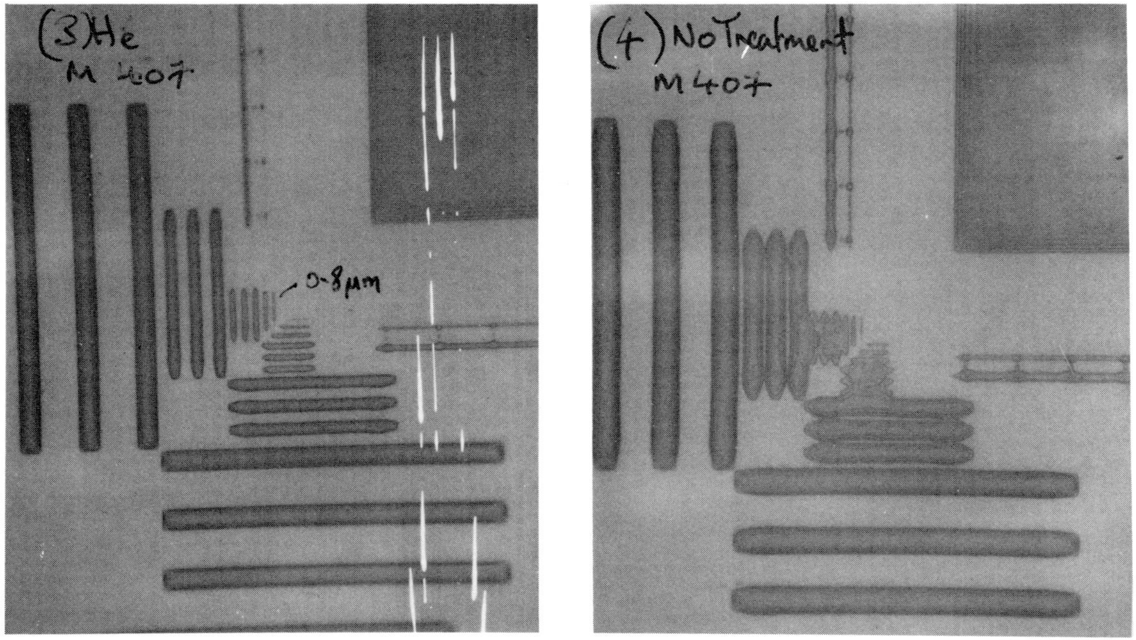

Fig 9 .17

Photos of plasma treatments using the barrel reactor of (1) CF_4 (2) N_2 (3) He (4) no treatment and all heated to 150 C for 10 min heating to the same extent as pre-treatment in a planar reactor. The CF_4 and He plasmas did reduce the flow occurring up to 150°C as shown in Fig. 9.17. Flow occurred on the control sample and the N_2 plasma pre-treated sample as also shown in Fig. 9.17. The reduction of flow after pre-treatment in CF_4 and He plasmas is probably due to the formation of a fluorinated skin over the resist in the CF_4 plasma and UV hardening for the He plasma. The N_2 plasma emits less UV than the He plasma and is chemically more inert to the resist film than the CF_4 plasma. This may be the reason why N_2 plasma pre-treatment failed to produce image stabilization.

The electrode potentials developed in a 317 reactor using an Ar discharge have been measured at various pressures in the previous chapter. Thus ion impact onto the resist surface may be as high as 300 volts at low pressure if the ions attain the full sheath voltage drop. However, this will only produce cross-linking at the surface, i.e. a 'skin effect'.

Infra-red spectra of plasma pre-treated resist films are shown in Fig. 9.18 and these show that little change has occurred in the bulk material. Also only a small amount of sensitizer has been destroyed after 20 min plasma exposure indicating a low UV output from the plasma above 300 nm wavelength. The generation of UV in this type of plasma by recombination events is low compared with the output of an UV lamp used in photolithography.

9.4.3 UV pre-treatments

Exposing a patterned photoresist film to deep UV radiation (254 nm) from a high pressure Hg Xenon source of 500 watts (as used on a Canon 500 mask aligner) produced only a marginal reduction in the resist etch rate. The exposures needed to be as long as 10 min to achieve any sig-

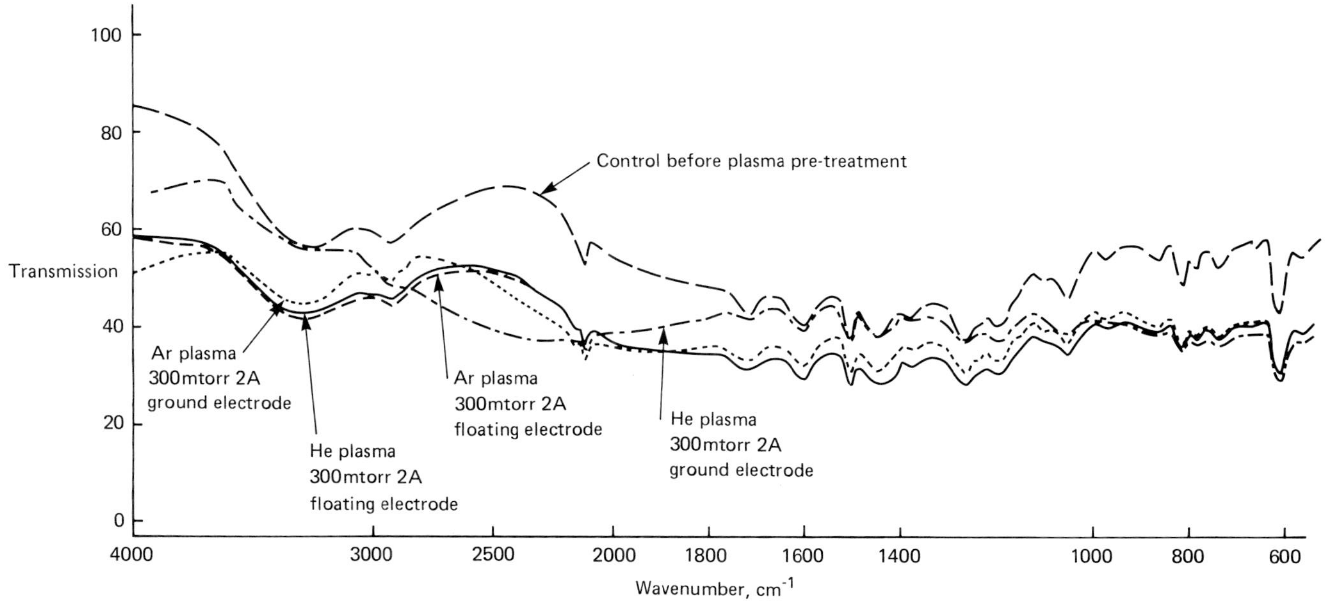

Fig. 9. 18

IR spectra of 2μm thick HPR 206 resist films before and after various plasma treatments in the Electrotech 317 Planar reactor. Al for 20 min pre-treatment time.

nificant difference. Short exposures to deep UV (< 5 min) appeared in some tests to increase the resist plasma etch rate slightly.

UV treatment can reduce flow at elevated temperatures as indicated in Fig. 9.19, which shows no flow on heating a patterned resist wafer to 180°C. However, some wrinkling occurred and this was also observed after heating the plasma pre-treated resists. The use of longer wavelength light sources (high pressure Hg) does not appear to reduce the flow properties (162).

Hiraoka and Pacansky (163) exposed photoresist samples (type AZ1350 and AZ2400) in vacuum to UV $\lambda < 300$ nm (Hanovia medium pressure Hg lamp) for 30 min and found that the resist does not flow up to 210°C. They believed that the sensitizer was activated by the UV to produce a ketene which normally reacts with water to produce an acid.

Fig 9.19

Deep UV (254 nm) exposed resist as a pre-treatment, then heated to 180° C for 1 hour.

However, in a vacuum the water can be desorbed from the resist film and thus the ketene reacts with the Novalac resin system producing some cross-linking. A repeat of this work by the writer has shown that there is no significant improvement in the reduction of the resist etch rate compared with that after UV exposure in air.

Infra-red spectra of a deep UV exposed resist sample for 10 min (atmospheric), given in Fig. 9.20, show that the absorption of the sensitizer C = N at 2120 cm^{-1} has almost disappeared. The C = 0 absorption at 1720 cm^{-1} due to acid formation has risen, whilst the C = 0 ketone (sensitizer) absorption at 1600 cm^{-1} has decreased. The small amount of sensitizer detected at 2120 cm^{-1} after the long exposure is because of the lower output of radiation above 300 nm wavelength from the deep UV lamp compared with standard UV Hg lamps.

Novalac polymers have a high absorbance $\lambda < 300$ nm and thus deep UV exposure will only be a skin effect. Poly methyl methacrylate (PMMA) resists have a high absorbance below $\lambda = 220$ nm and thus are useful positive resists for deep UV work at 254 nm but require long exposure for printing (168,67).

Another deep UV resist is Hunts WX159. This is a positive Novalac type resist containing a sensitizer which responds to 254 nm wavelength radiation and appears to be useful for good lithography using deep UV. Long exposures of deep UV (15 min) does reduce the resist erosion rate during subsequent plasma etching, albeit small.

9.4.4 Chemical pre-treatment

Experiments by the writer (67) have shown that formaldehyde CH_2O can be made to cross-link Novalac resists. However, this process also contaminates the wafer surface with the formation of paraformaldehyde $(CH_2O)_n$ which causes subsequent plasma process problems. The

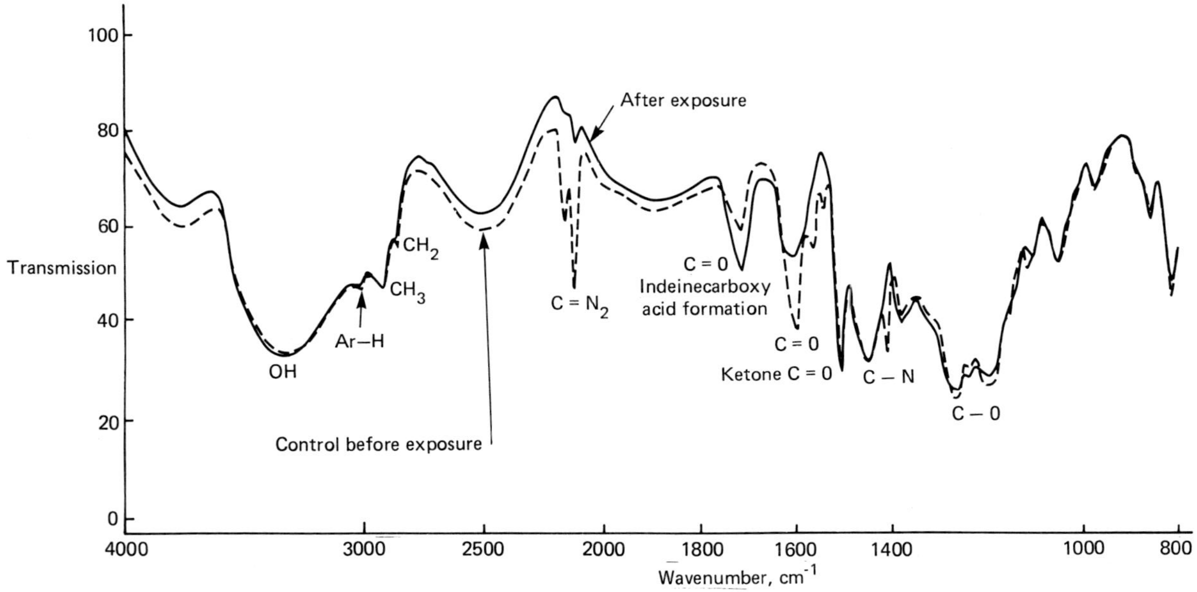

Fig. 9. 20

IR spectra of HPR 206 4μm thick resist film before and after exposure to 254nm deep UV radiation source for 10 mins.

requirements of a suitable chemical reagent for cross-linking Novalac resist are that it should have a low chemical hazard and will not introduce contamination to the wafer surface. In the hope of meeting these conditions and promoting cross-linking Luperox 101 was used. This is a dialkyl peroxide 2, 5 dimethyl, 2, 5 bis [t-butyl peroxide] hexane, which has the formula:

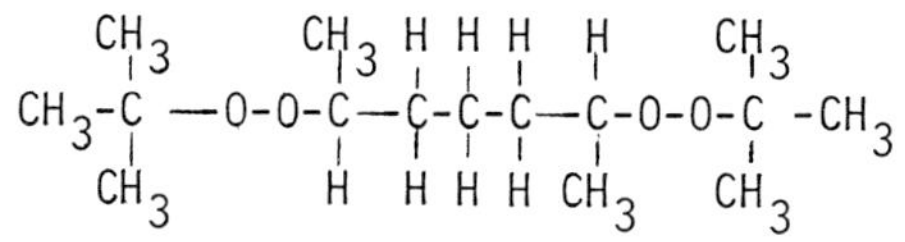

This compound is unstable in UV light and above 80°C. It is also a powerful oxidizing agent. When decomposed by heating or UV, it produces free radicals (bond rupture at the O-O linkage) which could produce cross-linking in the Novalac polymer. However, resist wafers which had been immersed in Luperox for up to one hour followed either by UV or heat treatment did not show a significant reduction in resist erosion during subsequent plasma etching. The resist swelled during resist pre-treatment showing that diffusion of Luperox or one of the decomposition products such as acetone had occurred. Experiments were ceased because the treatment was ineffective.

9.4.5 Ion impact pre-treatment

The resist patterned wafers were given to an experimental research implantation group at British Telecom Research Department. They placed the wafers into a Lintott series 3 ion implanter and exposed the wafers to an ion beam. The wafers were mechanically scanned across the ion beam (ion beam remains still). The beam used consisted of selected single charge Ar^+ ions. It was necessary when using a high energy ion beam to have a low dose rate by giving the wafers many passes across the ion beam to achieve a final high dose. This is to avoid the resist film becoming

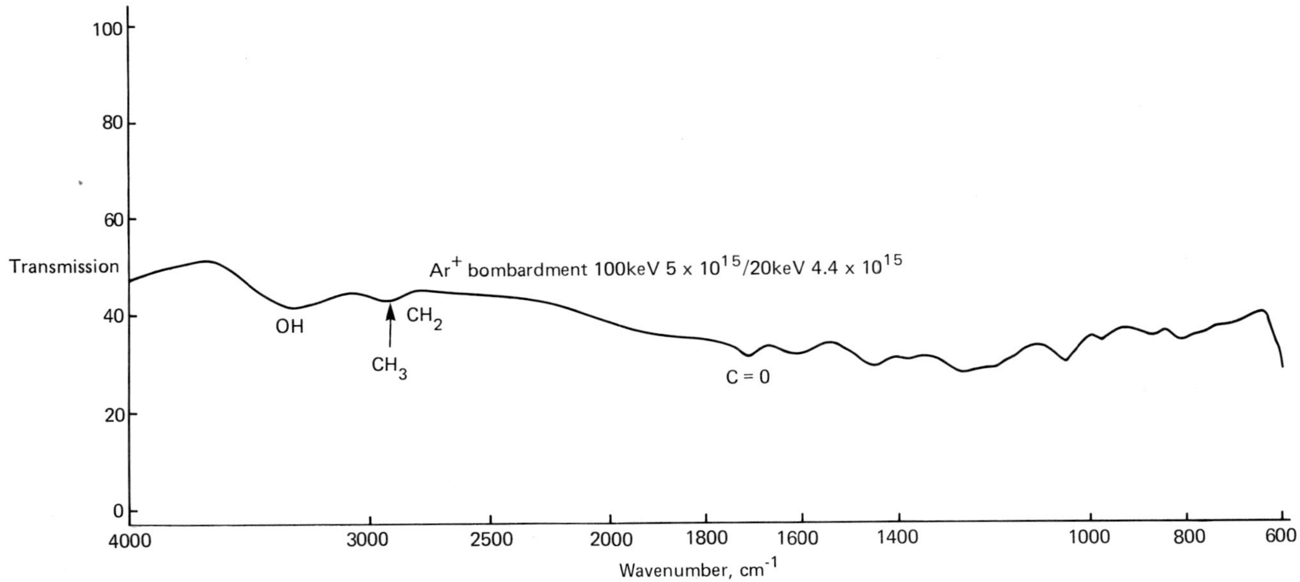

Fig. 9. 21

IR spectra 4.4 x 10^{15} dose at 20 keV and then dose 5 x 10^{15} at 100 keV Ar ions, using a Lintott implantation machine.

overheated, wrinkled and almost impossible to remove. The resist wafers were first given an ion Ar^+ dose of $4.4 \times 10^{15}\ cm^{-2}$ at an energy of 20 keV in one pass. This was followed by a total Ar^+ dose of $5 \times 10^{15}\ cm^{-2}$ at 100 keV energy in 20 passes.

The ion implantation process does significantly reduce the resist erosion rate during plasma etching processes. However, this as a resist pre-treatment procedure was impractical because throughput per wafer is low, the implanter time is expensive and the resist is often very difficult to remove after the plasma process.

Argon ions have a low penetration into resists, for example 100 keV Ar^+ ions penetrate $\sim$ 3000Å. On pre-treatment the resist film darkens probably due to carbon enrichment. Infra-red spectra, Fig. 9.21, of ion bombarded resists show loss of sensitizer and a reduction in the absorption of C-H, C = O and OH peaks.

Ions may be stopped by two mechanisms, firstly electronic, producing some ionization and secondary electrons and then, on slowing down, nuclear stopping where bond rupture and displacement can occur. It is probably the nuclear stopping by atom displacement of carbon in the polymer matrix which is contributing to the reduction in resist etch rate. However, the secondary electrons may cause cross-linking in the bulk of the film. As the ions only penetrate some 3000Å and the film develops a dark colour, it is believed that carbon enrichment has occurred at the surface.

9.4.6 Electron beam pre-treatment

This section describes the use of a 10 keV scanned electron beam for direct impact induced bond scission in resist. The electron beam was employed to cause cross-linking in the resist layer to produce a structure more resistant to plasma erosion as discussed in the introduction.

The patterned resist wafers were divided into three experimental groups and separately exposed to the 10 keV scanned electron beam in an annealing machine (Lintech SEZA), which gave a low and stable beam current at 10 keV. The focussed electron beam had a spot size of 250 μm and was scanned at a line and frame frequency of 20 kHz and 100 Hz respectively. The 10 keV electrons were estimated to penetrate 1.6 to 1.9 μm into the resist.

Three sets of wafers were treated. In the first experiment they were irradiated at 10 keV with a total integrated charge dose of between 3 mC cm^{-2} to 10 mC cm^{-2}. During this irradiation a dose rate of 0.01 mC $cm^{-2}s^{-1}$ was used and the resist wafers rested on a copper heat sink to maintain the wafer temperature below the glass transition of the resist. After irradiation the wafers were removed from the annealing machine and heated on a hot plate at temperatures between 130°C and 200°C. No flow or bevelling was observed in any of the resist patterned wafers, i.e. the original resist profile was maintained.

This experiment was repeated, but this time the wafers, during irradiation, were thermally isolated by resting them on three small silicon pins. This ensures the temperature of the wafers rose during electron irradiation and could be controlled via the dose rate (164). For example, for a dose of 5 mC cm^{-2} the wafer temperature ranged from 80°C (dose rate of 0.011 mC $cm^{-2}s^{-1}$) to 200°C (0.034 mC $cm^{-2}s^{-1}$). The SEM photographs of resist before and after electron beam irradiation using thermal isolation are shown in Fig. 9.22. Flow was not observed in this experiment or after any of the other irradiation tests.

(1)

(2)

Fig 9.22

SEM photographs (1) 3 µm line, 2 µm thick HPR206 before electron beam irradiation (2) 2 µm line after electron beam irradiation 10 KeV, 5 mC cm^{-2} dose/temp rise 200° C). No flow observed.

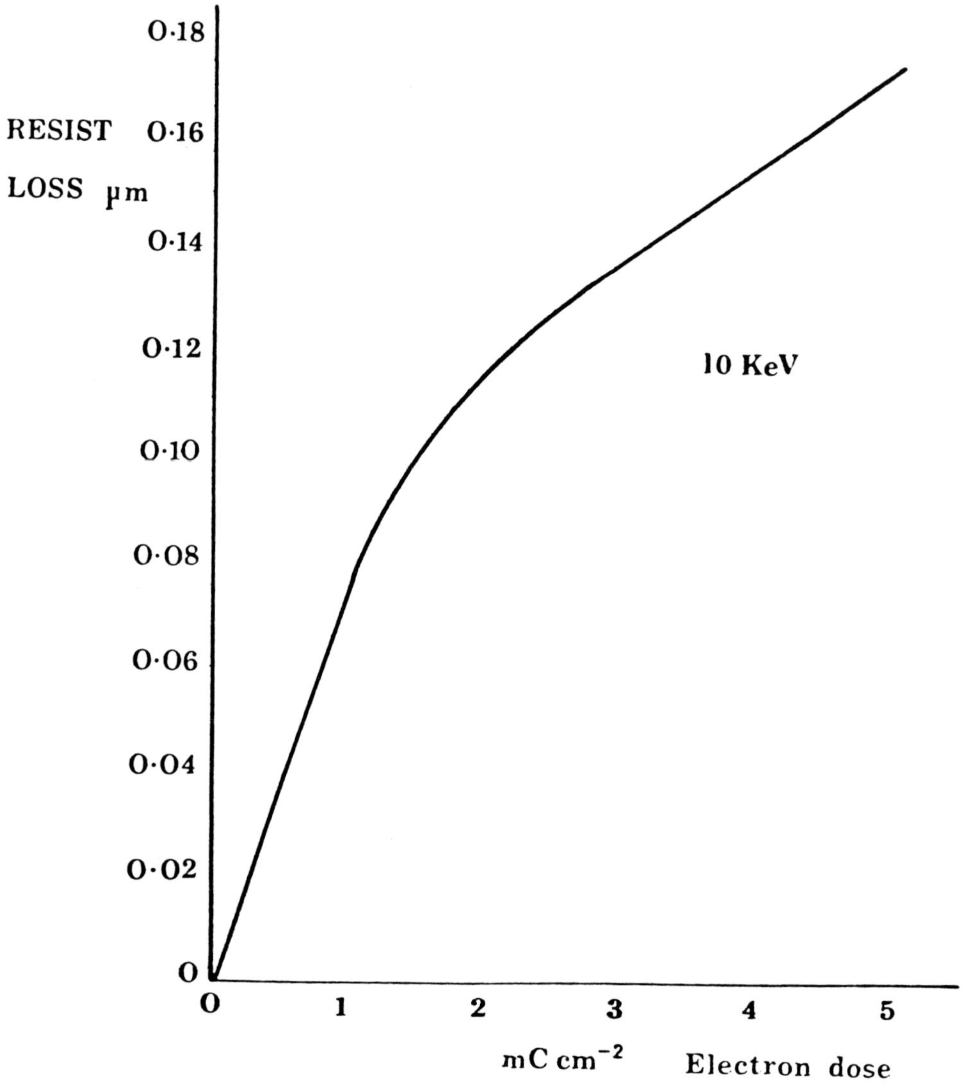

Fig 9.23

Plot of resist thickness loss during irradiation at various dose levels (10 keV)

Resist patterns probably do not flow during electron irradiation because the rate of cross-linking is fast compared with the rate of temperature rise. A small amount of cross-linking can prevent plastic flow or glass transition (165).

The second experiment was done at a fixed dose rate of 0.24 mC $cm^{-2}s^{-1}$ and with total integrated charge doses of 0.5 mC cm^{-2} to 5 mC cm^{-2} (in thermal isolation). The resist thickness loss during irradiation pre-treatment was measured and the results are shown in Fig. 9.23. The pre-treated resist wafers with a heat treated (120°C, 15 min) resist wafer as a control were plasma etched. The resist thickness loss when plasma etching aluminium and polysilicon substrates for various doses during pre-treatment are shown in Fig. 9.24. The SEM photographs in Fig. 9.25 compare (a) heat treated control and (b) the electron beam pre-treated resist (5 mC cm^{-2}) both on polysilicon substrates which were plasma etched at the same time. The SEM photograph of the heat treated control which is the standard treatment given to resist prior to plasma etching processes shows little resist left and a bevelled profile which has caused a bevell in the polysilicon and loss of linewidth. The SEM photograph of the electron beam pre-treated resist shows straight edges with a large amount of resist left.

The electron beam pre-treatment only produces a significant reduction in resist etch rate above doses of 3 mC cm^{-2}, as shown in Fig. 9.24. This also shows that there is small increase in resist etch rate at doses below 2 mC cm^{-2}. Fig. 9.23 shows a change in slope for the resist loss during pre-treatment, also at about 2 mC cm^{-2} dose.

In the third experiment the dose level was kept constant at 5 mC cm^{-2} and the dose rate to the wafers in thermal isolation was varied from 0.1 to 0.5 mC $cm^{-2}s^{-1}$. The resist loss from electron irradiation pre-treatment at various dose rates is shown in Fig. 9.26. On plasma

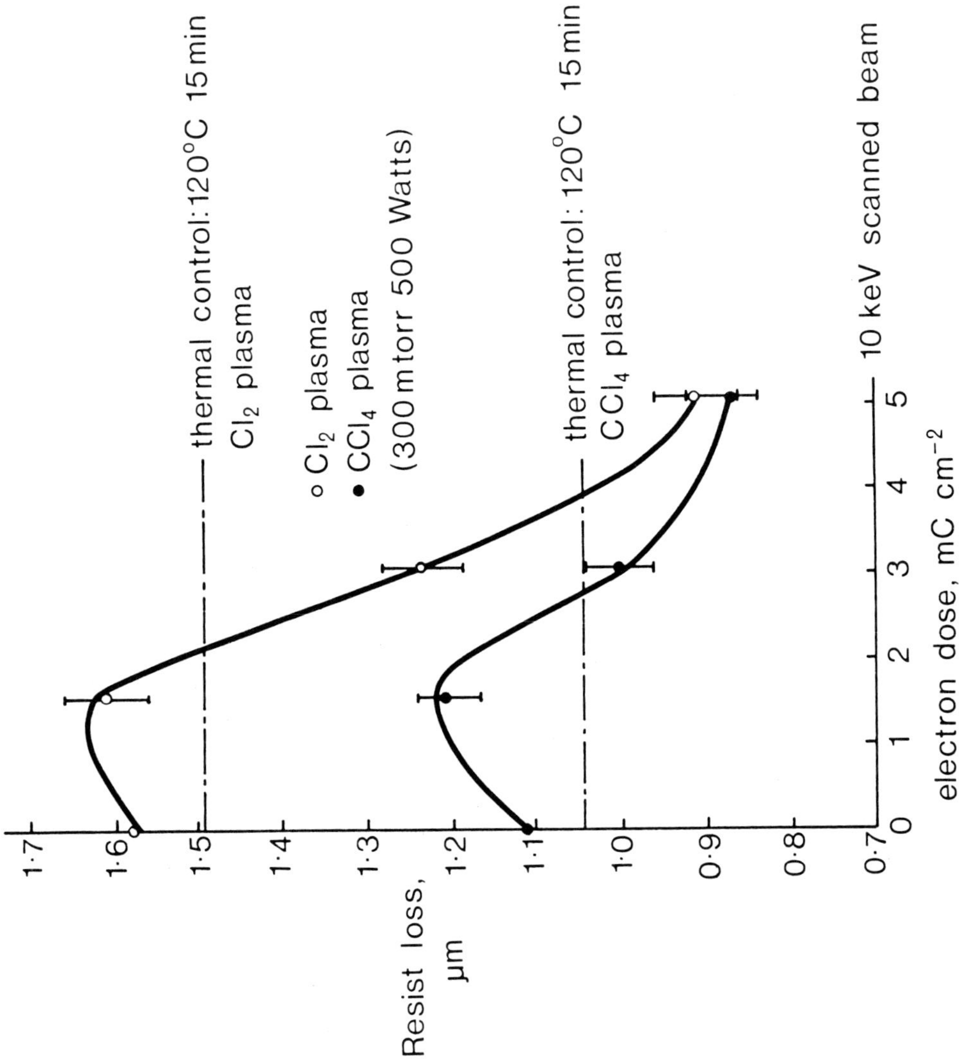

Resist loss, µm
1·7
1·6
1·5
1·4
1·3
1·2
1·1
1·0
0·9
0·8
0·7
0
1
2
3
4
5
electron dose, mC cm^{-2}
thermal control: 120°C 15 min Cl_2 plasma
thermal control: 120°C 15 min CCl_4 plasma
○ Cl_2 plasma
● CCl_4 plasma
(300 mtorr 500 Watts)
10 keV scanned beam

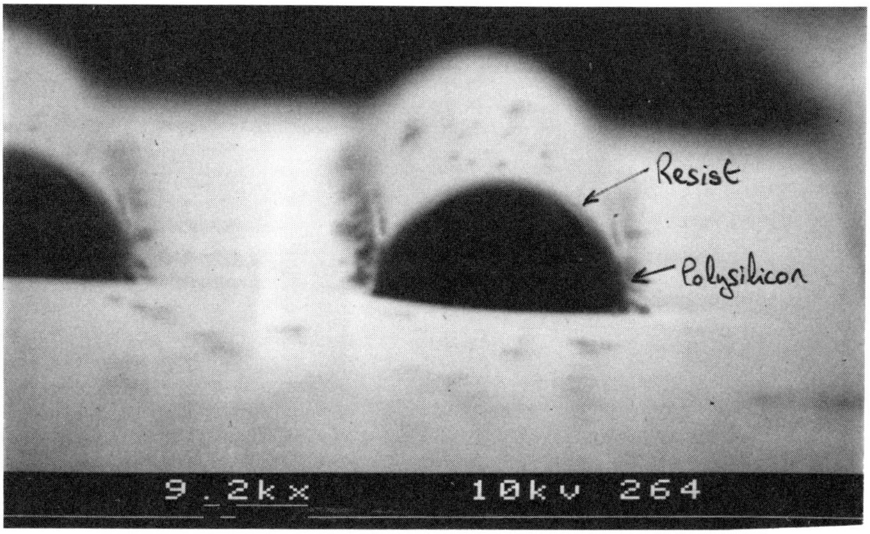

Fig. 9.25 (1). Heat treated resist HPR206 for 120°C 15 min. Then etched in a Cl_2-plasma, 300 mtorr, 2 amps for 10 min.

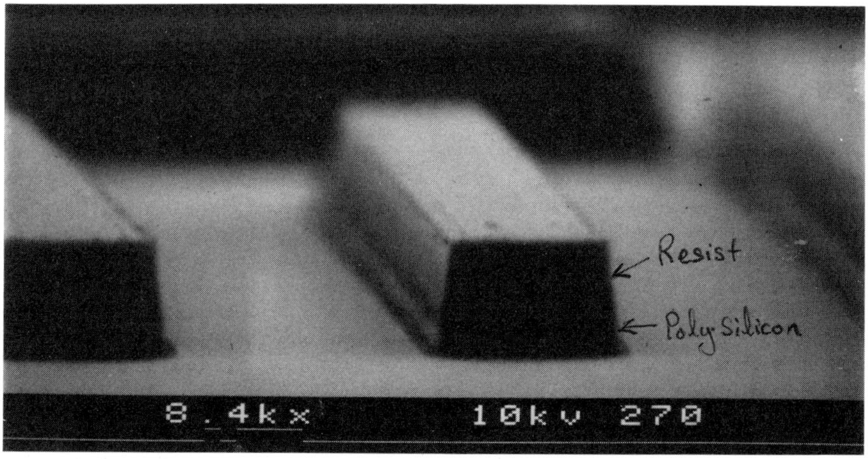

Fig. 9.25 (2). Electron beam pre-treated HPR206 resist at 5 mC cm^{-2}, 10 keV. Then etched in a Cl_2-plasma, 300 mtorr, 2 amps for 10 min.

Fig 9.24

Plot of resist loss during the etching of 1 μm of Al in CCl_4 plasma and 1 μm of polysilicon in Cl_2 plasma at various dose levels (10 keV)

etching the resist wafers with the same dose but at higher dose rates, a further reduction in resist erosion rate was observed. However, at high dose rates (0.5 mC $cm^{-2}s^{-1}$) there is more resist loss during pre-treatment and the wafer temperature can exceed 250°C after which the resist is not removed in fuming nitric acid. Resist wafers (2 μm thick resist) that did not exceed 200°C during the electron irradiation pre-treatment were removed in fuming nitric acid probably because at the resist/substrate interface the resist is not highly cross-linked.

The lowering of the resist etch rate by up to 40% from electron beam pre-treatment at high doses > 3 m cm^{-2} indicates a high degree of cross-linking and structural changes have occurred.

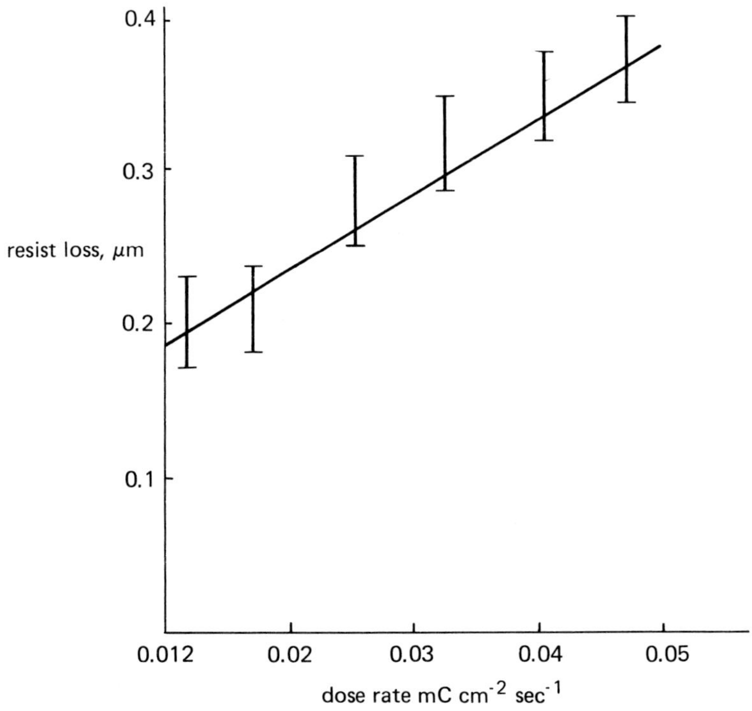

Fig. 9. 26

Plot of resist loss during electron beam irradiation 5mC cm^{-2} (10keV) constant dose vs dose rate

The IR spectra of electron irradiated Novalac resist films shows that there is a large loss of oxygen at a dose of 5 mC cm^{-2}. The oxygen loss was that in the form of C = O and OH. There was also a loss of -CH. There appears to be a general decrease in all H-C and O-C absorptions. However, there may be a small increase in C-C absorption but with small dipole changes it is difficult to detect by IR spectrometry. However, the samples become darker with larger doses and this would arise from the relative carbon content rising. Figs. 9.27 and 9.28 are IR spectra of (HPR206) Novalac resists that have been electron bombarded with 5 mC cm^{-2} and 1 mC cm^{-2} doses respectively with an untreated resist for comparison. Figs. 9.29 and 9.30 are IR spectra of resist films after pre-treatment with the same dose of 5 mC cm^{-2} but at different dose rates of 0.13 and 0.47 mC $cm^{-2}s^{-1}$ respectively. The results show in Figs. 9.29 and 9.30 that there is no obvious difference in the final beam effect.

The large reduction in resist erosion rate during irradiation appears when the OH (around 3300 cm^{-1}) and C = O (at 1720 cm^{-1}) peaks start to reduce in the IR spectra, between 3 to 5 mC cm^{-2}. Wafers treated at 10 mC cm^{-2} had a marginally lower resist erosion rate than the 5 mC cm^{-2} treated wafer.

Obviously the precise mechanism of cross-linking is unknown because of its complexity. However, when electrons at a primary energy of 10 keV strike the film they release secondary electrons whose energy distribution has an average value of about 5 eV. Apart from the incident high energy electrons the low energy secondaries can excite and disrupt chemical bonds in the polymer and sensitizer. By this means, radicals are formed. The radicals produced can undergo several reactions. One type is an elimination reaction releasing molecules such as H_2O and CO and then form an unsaturated molecule which can later cross-link.

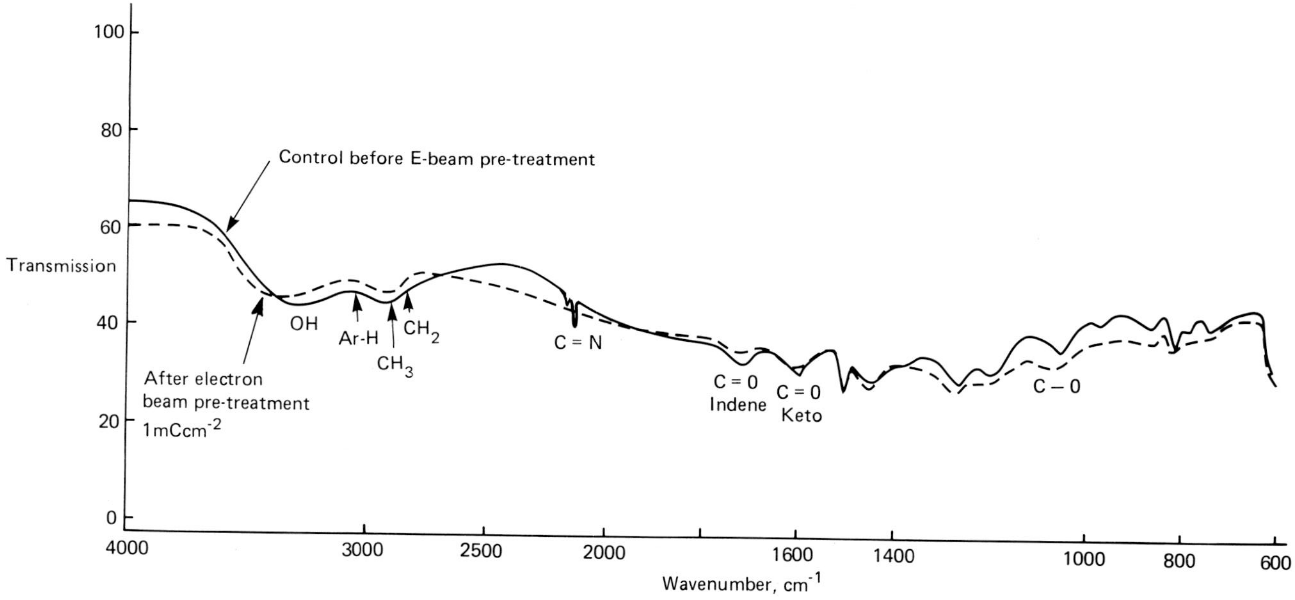

Fig. 9. 27

IR spectra of an electron irradiated photoresist 2μm thick printed HPR 206 1 mC cm^{-2}, 10 keV at a dose rate of 0.018 mC cm^{-2} s^{-1} with control for comparison.

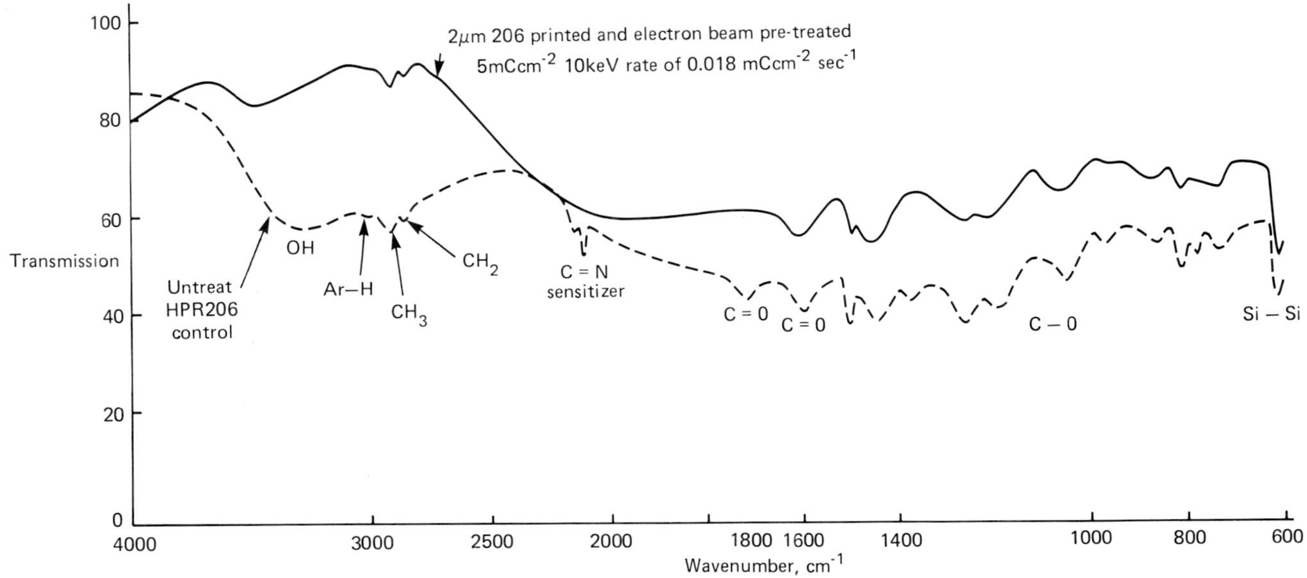

Fig. 9. 28

IR spectra of an electron irradiated photoresist at 5 mC cm^{-2} 10 keV with control for comparison.

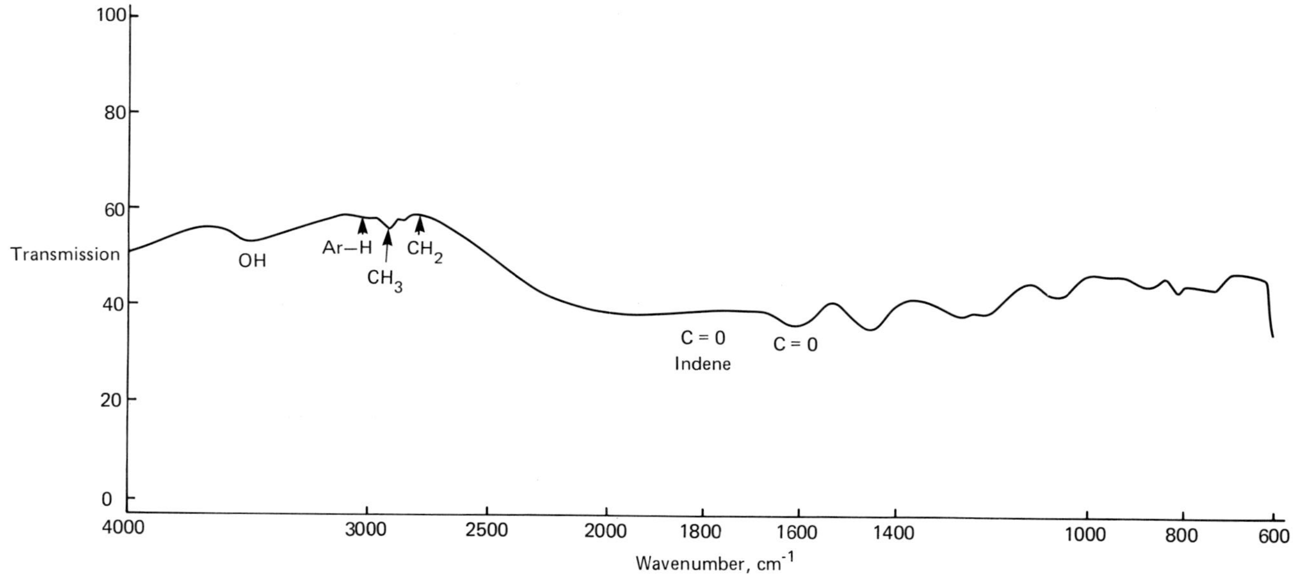

Fig. 9. 29

IR spectra of an electron irradiated photoresist at 5 mC cm^{-2} 10 keV at a dose rate of 0.13 mC cm^{-2} s^{-1}.

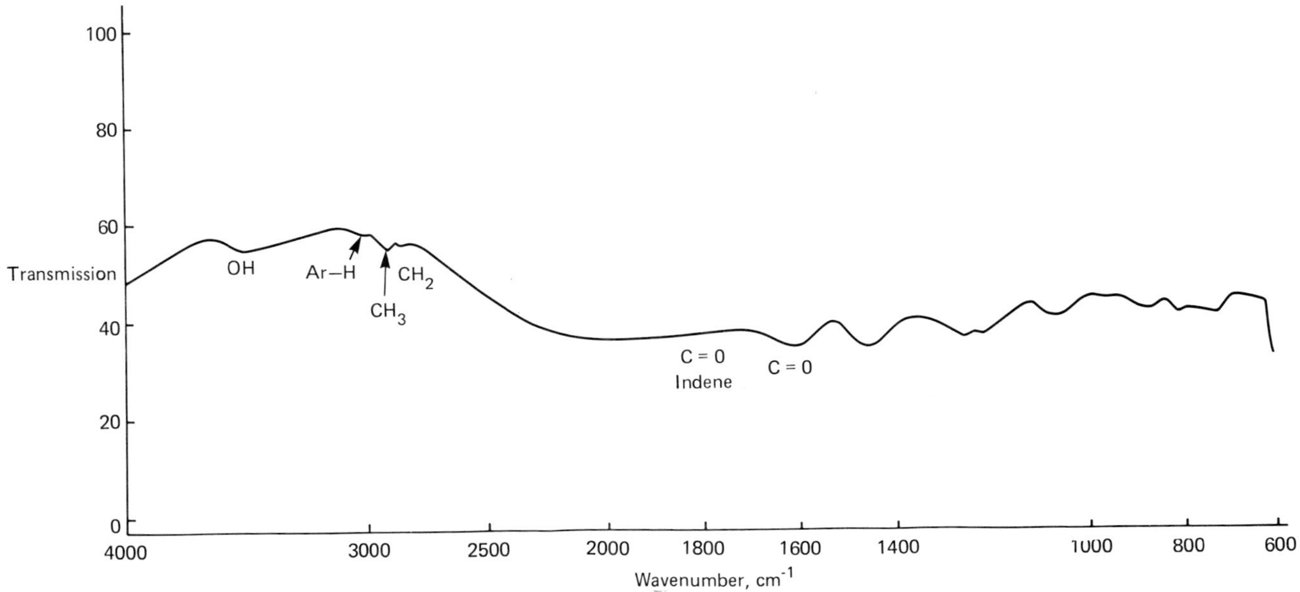

Fig. 9. 30

IR spectra of an electron irradiated photoresist at 5 mC cm^{-2} 10 keV at a dose rate of 0.47 mC cm^{-2} s^{-1}.

On electron bombardment of less complex resists such as PMMA there are large resist pre-treatment losses (50% of the original film thickness at 5 mC cm^{-2}) because of the large oxygen content in the resist. Fig. 9.32(a) is an IR spectra of PMMA before and after electron irradiation (5 mC cm^{-2}, 10 keV) showing the large loss of C = O in the irradiated spectra. The remaining PMMA resist after the 5 mC cm^{-2} electron dose (10 keV) does have a lower plasma erosion rate although some flow was observed, which indicates a slower rate of cross-linking than in Novalac

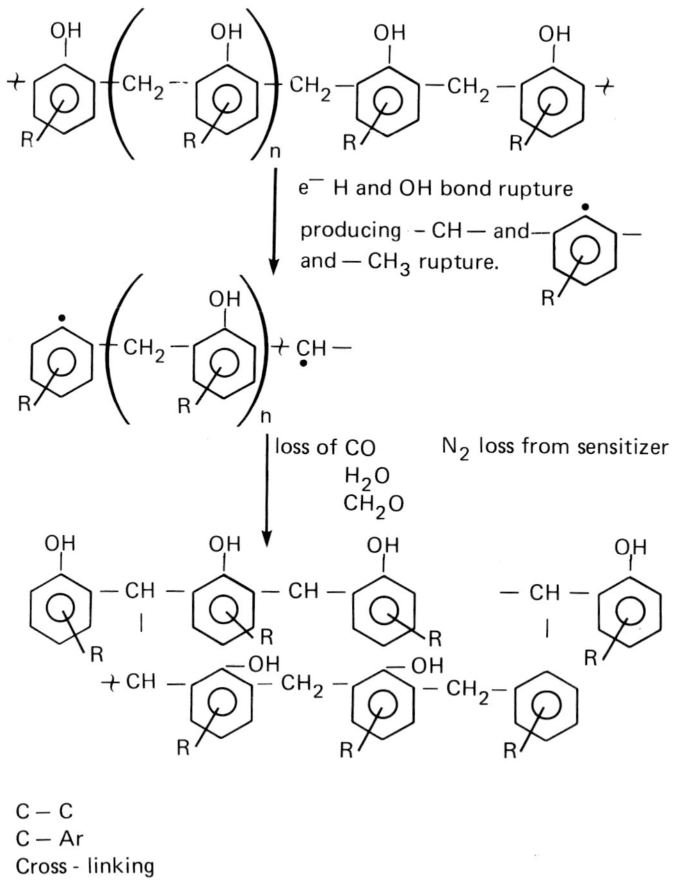

Fig. 9. 31

Proposed mechanism for cross linking by electron irradiation for Novolak polymers.

polymers. The reduction in resist etch rate in plasmas occurs at about 5 mC cm^{-2}, when the C = O absorption has disappeared from the IR spectra as shown in Fig. 9.32(a). The proposed reactions for cross-linking during electron irradiation of Novalac and PMMA resists are shown in Figs. 9.31 and 9.32(b) respectively.

Workers (174) have reported that during electron irradiation at low doses (30 μC cm^{-2}, 20 keV), PMMA undergoes main chain scission reactions thus lowering the average molecular weight. The areas irradiated can be preferentially dissolved away in a suitable solvent and hence PMMA is a good positive electron beam resist. However, the work here has shown that PMMA had also cross-linking events which occur at high electron dose levels (5 mC cm^{-2}) with large pre-treatment thickness loss because of the loss of CO, CO_2 and H_2O from the PMMA.

9.5 Miscellaneous Resists

A comparison of the etch rates of a number of different resists with no pre-treatment, has been made. They are listed in order of increasing etch rate in plasmas; HUNTS WX159 < HUNT HPR206 < Macdermid PR74 < PMMA. Although the differences are small during Cl_2 plasma etching they correlate with a decrease in aromatic nature from WX159 which is highly aromatic to PMMA which has no aromatic characteristics and the highest etch rate, as shown in Fig. 9.33. Thus, the higher the aromatic characteristic the lower will be the resist etch rate (166).

Fig. 9.33 also includes a measurement of resist etch rate of the highest aromatic resist available which was a polystyrene based polymer used as a high resolution negative working electron resist. The resist is known as HRN and is made by Philips.

The difference in resist etch rate between PMMA and WX159 is greater in CCl_4 plasmas with PMMA etching $\sim 2\frac{1}{2}$ times faster than WX159.

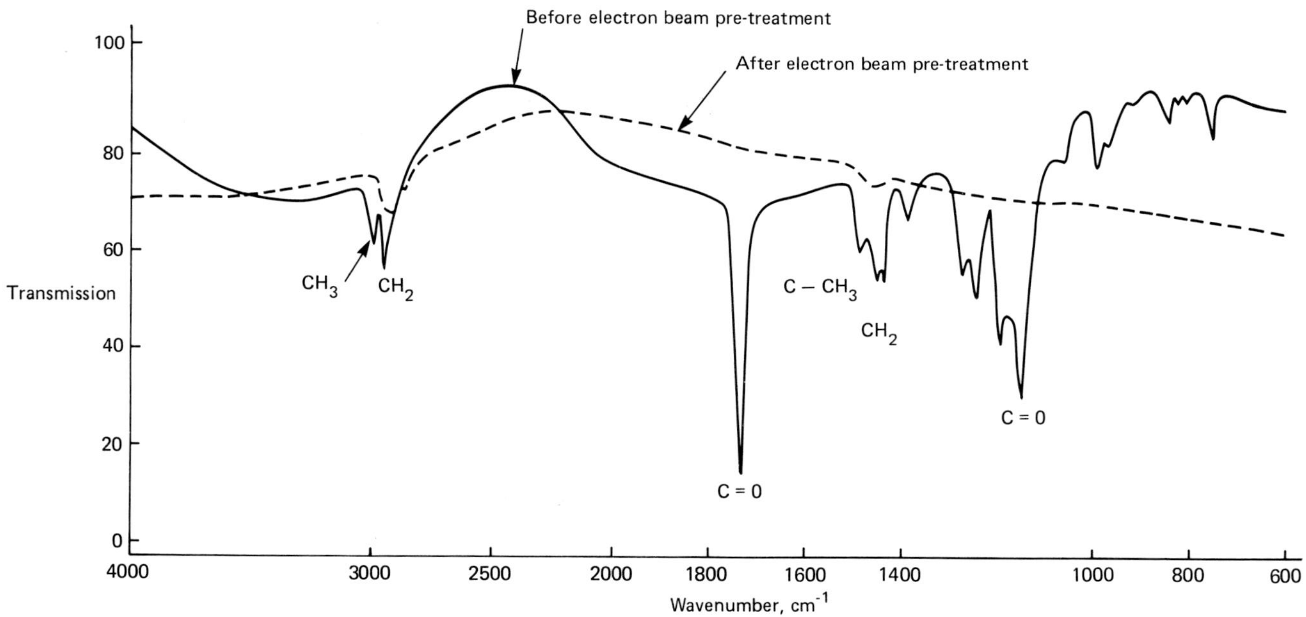

Fig. 9. 32a

Comparison between untreated PMMA 3μm thick and after electron beam pre-treatment at 10keV 5MC cm^{-2}.

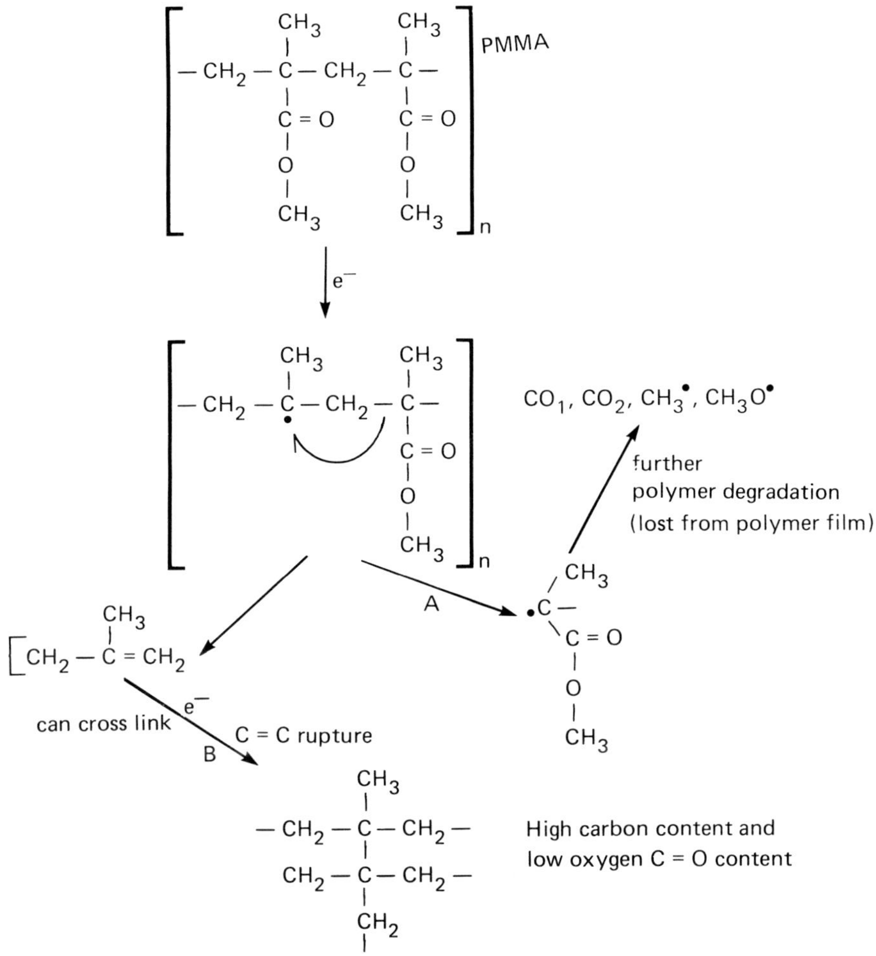

Fig. 9. 32b

Proposed mechanism for electron induced bond rupture in PMMA. Homolysis of main chain C — C as first step. Also acylcarbon — oxygen σ bond homolysis and rapid decarbonglation leading to the same products. Reaction B ultimately leads to cross-linking.

Some resists respond better to a particular pre-treatment than others. Macdermid resists responds very well to electron irradiation pre-treatment, with a reduction in plasma resist etch rate at low doses, i.e. less than 0.5 mC cm^{-2}. PMMA and WX159 appear to respond better to UV treatments compared with Macdermid and HPR206 probably since they are

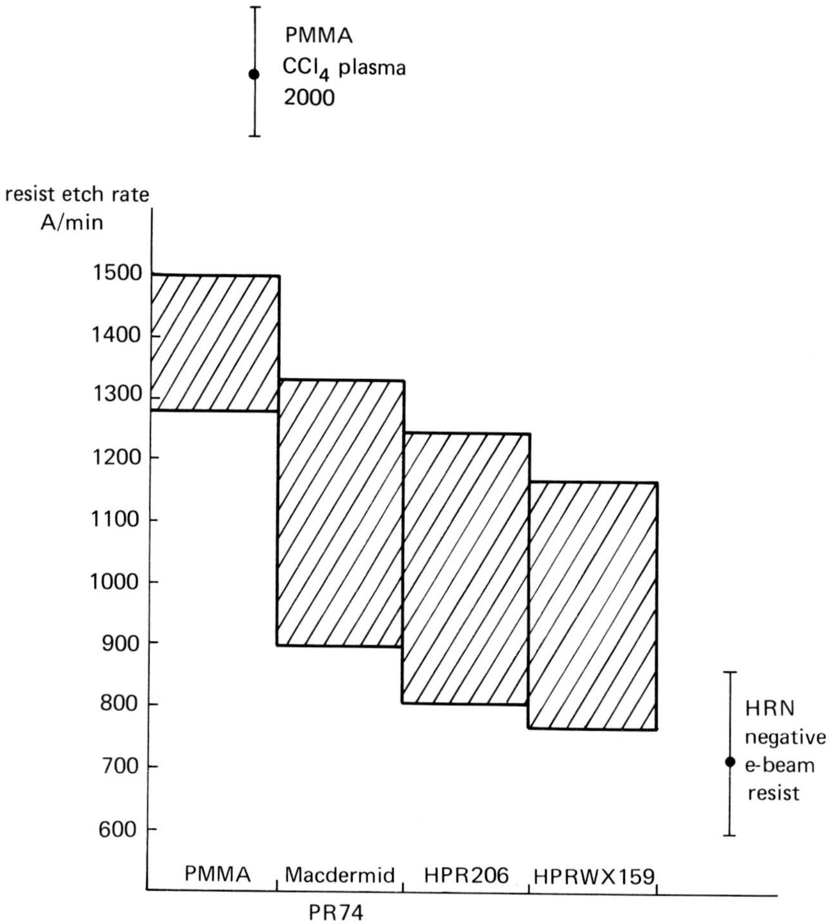

Fig. 9. 33

Plasma etch rate vs type of resist untreated using a Cl_2 plasma.

deep UV resists, allowing the UV to penetrate. The results of WX159, PMMA and Macdermid to some of the treatments are shown in Fig. 9.34.

9.6 Comparison of Resist Pre-Treatment Techniques and Summary

Table 9.2 is a summary of the resist pre-treatment conditions, the results and comments. Table 9.3 summarises the plasma etching results in terms of selectivity, i.e. substrate etch rate and pre-treatment resist etch rate as a ratio.

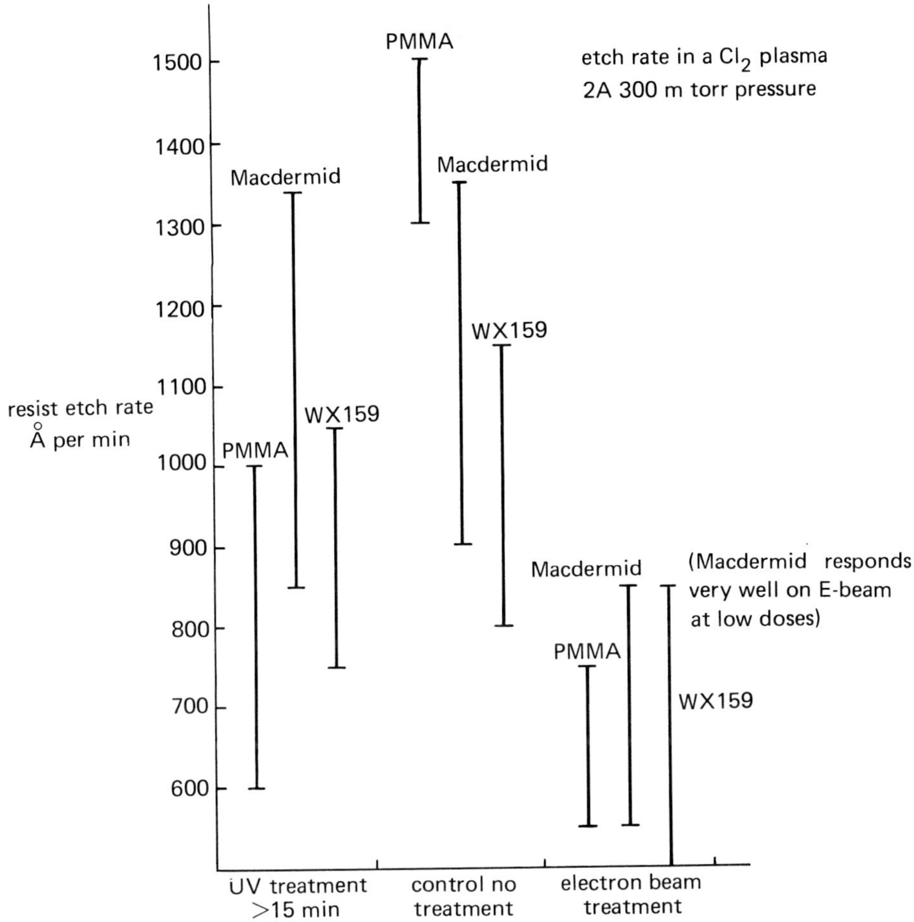

Fig. 9. 34

Plasma etch rates as a function of pre-treatment for WX159, PMMA and Macdermid PR 74 resists.

The foregoing research shows that the largest reduction of resist etch rate in a plasma is achieved by electron irradiation and high energy Ar^+ bombardment pre-treatments. However Ar^+ bombardment is not considered usable because of the long wafer pre-treatment time of up to five hours. The electron irradiation pre-treatment produces rapid cross-linking in the resist with no flow and almost doubles the selectivity between etched

TABLE 9.2

APPENDIX FOR PRE-TREATMENT

Pre-treatment Process	Resist Results and Comments
Heat pre-treatment IR lamps and hot plates between 100° C and 250° C, for ½ hr.	Flow of resist occurs above 120° C as shown in Fig 9.8. Thus for accurate measurements samples were prepared using the trilevel process shown in Fig 9.10b. Fig 9.11 is a plot of resist etch rate vs, temperature treatment a Cl_4 plasma. It shows that there is only a significant improvement in resist etch rate for pre-treatments above 200° C, but above this temperature the resist becomes very difficult to remove by wet methods.
Ultra Violet pre-treatment 254 mm high pressure Hg/Xe source 500 watts fitted to a Cannon 500 mask aligner 10 min. exposure time.	Only a marginal reduction in resist etch rate was observed. However good image stabilization ie no flow, was observed at temperatures of 180° C, 1 hr. This is shown in Fig 9.19 but the resist surface was wrinkled.
Chemical pre-treatment Immersion time up to 1 hr using Luperox 101 Activated by UV exposure or heat to 80° C.	This reagent was used as it is one of the safest organic peroxides. However no reduction in resist etch rate was observed. No image stabilization was observed.
Ion Implantation Ar Ion bombardment in to photoresist films Dose: (1) 4.4×10^{15} cm^{-2} at 20 keV followed by 20 passes at 5.0×10^{15} cm^{-2} at 100 keV	There was observed a significant reduction in resist etch rate. However ions must have a low dose rate to prevent the resist from becoming hot and wrinkling, thereby becoming difficult to remove. For this reason the process requires a long pre-treatment time of 5 hr per wafer and this is not considered as a practical method.
Electron Beam Treatment Scanned electron beam bombardment in to photoresist, at 10 keV with a dose between 1 and 5 m C cm^{-2} using a Lintech SEZA.	Resist films after bombardment have darkened above 3 m C cm^{-2} and there is a significant reduction in resist etch rate; shown in Fig 9.24. The films do not appear to suffer from wrinkling or flow even when taken up to 200° C (as shown in Fig 9.22). The method has a pre-treatment time of 7 min per wafer and therefore can be a practical technique.

Plasma pre-treatment a. Planar reactor (317) Ar or He plasma, 100 and 300 mtorr, 2 amps For 10 min.	There is a small reduction in resist etch rate. However there is a large resist loss on resist pre-treatment. Floating the wafer during plasma treatment does not reduce the resist etch rate significantly but does reduce the pre-treatment loss. The large run to run variation in measurements of pre-treatment losses may be due to chamber contamination. Fig 9.16 shows a patterned area of resist which has been plasma pre-treated, and heated to 180°C for 1 hr with no flow occurring.
b. Barrel reactor (IPC) with sheild, CF_4, N_2 and He plasmas 700 mtorr, 400 watts, 15 min	No reduction in resist etch rate. There was a reduction in flow up to 150°C for CF_4 and He plasma shown in Fig 9.17.

TABLE 9.3

SELECTIVITY BETWEEN AL CONTAINING ~ 2% Si AND PRE-TREATED HPR206 RESIST (a)
" " Poly Si " " " (b)

Pre-treatment	Al/Si 2% in CCl_4 plasma	Polysilicon in Cl_2 plasma
Control (no pre-treatment)	3.2:1	1.4:1
Heat (100-150°C)	3.2:1	1.45:1
Deep UV	3.2:1	1.55:1
Luperox (Organic peroxide)	3.2:1	1.4:1
Heat (TR1 level 200°C)	3.4:1	1.55:1
Plasma	3.6:1	1.8:1
Electron beam	5:1	3.1
Ion implantation	7:1	3.7:1
	Al:Resist	Si:Resist

substrate and resist mask. For doses above 3 mC cm^{-2}, the pre-treatment time is about 7 min per wafer and is therefore a practical method of resist hardening. By reducing the resist erosion rate during plasma etching, more process flexibility is allowed,such as over etching.

Untreated aromatic polymers such as HPR206, plasma etch slower than non-aromatic polymers, such as untreated PMMA.

Large numbers of free radicals are produced in the plasma at the conditions used for etching, i.e. 300 mtorr pressure, 2 Amps r.m.s. However, resist HPR206 is not significantly etched by free radicals alone during a Cl_2-plasma etching process (67). One possible reason could be attributed to some wafer cooling from the ground water cooled electrode. However, with the addition of Cl^+ and Cl_2^+ bombardment from the plasma, rapid etching of the resist occurs, i.e. ion activated. By initially cross-linking the resist its subsequent erosion in a planar reactor can be reduced because impacting plasma ions must rupture several bonds in order to volatilise or remove small fragments of resist. Bonds once damaged by ion impact from the plasma can react with the available free radicals and in some cases chemisorb gas molecules near the vicinity of impact aiding resist volatilisation.

If the ion bombardment energy increases in a plasma to a value where the ion can rupture several bonds per impact then the resist hardening methods will become less effective. This has been observed by the writer using a low pressure experimental RIE system where ion energies of 1500 volts were reached. This was because of the high electrode potential and the low pressure which allowed ions to transverse the sheath with no collisions and hence achieve full sheath potential.

Voltages developed in the planar reactor used in this study was one subject discussed in the previous chapter but chlorine gas was not available at the time measurements were made. Recent measurements made by the same method described in the previous chapter for a Cl_2-plasma gave $V_{r.m.s.}$ across the ground electrode sheath to be ∿ 250 volts at 300 mtorr 2 Amps r.m.s. In addition the calculations used in Section 8.3.5 applied to a Cl_2 plasma (300 mtorr 2 Amp r.m.s.) reveal that the untreated HPR206 resist etch rate exceeds the ion arrival rate indicating that some ions must rupture more than one bond per impact. By cross-linking HPR206 resist with electron irradiation at 10 keV energy, 5 mC cm^{-2} dose the resist etch rate is almost halved in the Cl_2-plasma etching process and is because there is a lowering of the bond rupture rate by ion impact.

A method is needed to measure both qualitative and quantitative cross-linking processes occurring through resist layers from pre-treatments that are bulk or skin effects. One method being considered by the writer is Photon Acoustic Spectroscopy (173). This method could help understand still further the amount of cross-linking, the reaction mechanisms and its contribution in reducing plasma erosion through the resist film layer. This will enable a resist polymer to be 'tailored' for a particular plasma process. One example would be to measure the extent of cross-linking from a lower energy electron irradiation pre-treatment through the resist, so that thin resist layers (0.5 μm) may be hardened and removed after plasma etching.

In the study here the amount of cross-linking in the bulk of the resist film has not been directly measured for resist pre-treatments. However, the physical and chemical changes that have occurred in the resist during ion and electron irradiation pre-treatments are believed to cause a high degree of cross-linking and carbon enrichment. This enables the selectivity between aluminium or polysilicon to resist during etching to be doubled. However this improvement in selectivity for some processes is insufficient and an alternative resist technique is still needed.

9.7 Further Work

By growing a silica film on polysilicon, patterning with resist and etching the exposed oxide produces a silica mask which has a very low plasma etch rate in a Cl_2-plasma. This gives a selectivity between silica mask to polysilicon in a Cl_2 plasma of about 1:40, but requires the extra process steps. Recent work by the writer (patent applied for) has introduced an Si-O cross-linked matrix into the Novalac polymer by silylation of the non-sterically hindered OH groups. The general principles are shown in Fig. 9.35. The Si-O matrix in the Novalac polymer

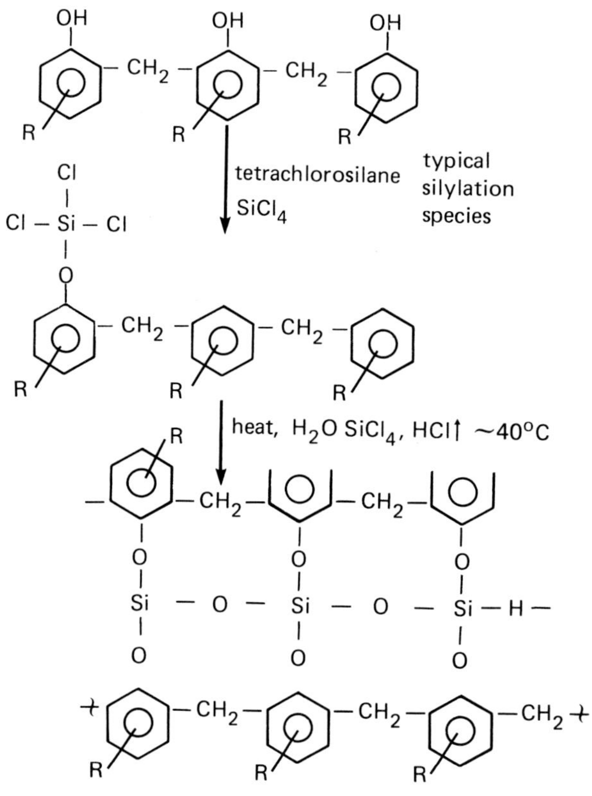

Silylation can produce Si — O matrix which is a good barrier for a Cl_2 and O_2 plasmas

Fig. 9. 35

Proposed Silylation of a Novolak polymer.

reduces the resist etch in O_2 and Cl_2 plasmas and possibly CF_4 (further work needed).

Other polymers such as polyimide used as a dielectric layer in semiconductor fabrication when fully cured has no O-H or N-H groups present, i.e. acidic hydrogen available for silylation. By silylation of a patterned Novalac resist mask over polyimide, high selectivity can be obtained between mask and polyimide in an O_2-plasma because of Si-O formation. An alternative is to develop photoresist systems for specific plasma processes such as poly trimethylsilyl styrene and polymethylcyclosiloxane which can form an SiO_2 layer and which appears to be resistant to O_2 and Cl_2 plasmas.

Workers (167) believe that some types of polymers can consume the energy of plasma excited species in the process of cross-linking, thereby reducing the etch rate. Another alternative to produce masks of low etch rate in a plasma is to develop inorganic optical resists (157).

REFERENCES AND NOTES

1. L. Holland, Surface Technology 11 (1980), 145.
2. R.L. Bersin, Solid State Tech., April 1978, p.117.
3. R.G. Poulson, J. Vacuum Sci. and Tech., Vol. 14, No.1, 1977.
4. G.N. Jackson, Thin Solid Films 5 (1970), 209-249.
5. D. Bollinger and R. Fink, Solid State Tech., Nov. 1980, p.79.
6. J.M. Moran and D. Maydan, Bell System Tech. J., 58, p.1027, 1979.
7. A.R. Reinberg, U.S. Patent 3,757,733.
8. R.A. Heinecke, Solid State Electronics, Vol. 18, p.1146, 1975.
9. U. Winkler, Report from Valvo, West Germany.

10. M.L. Hitchman and V. Eichenberger, J. Vac. Sci. and Tech., Vol. 17, No.6, Nov./Dec. 1980.

11. R.C. Booth, BTL Private Communication.

12. C. Tokauaga and D.W. Hess, J. Electrochem., SST, April, p.928, 1980.

13. Consumption of Cl· in a CCl_4 plasma by reaction with an aluminium electrode will reduce the Cl· concentration, this will increase the $CCl_3^{\cdot}$ concentration by the reduction in Cl· + $CCl_3^{\cdot}$ recombination events. The increase in C:Cl ratio in the system consequent on the [$CCl_x^{\cdot}$] increase will enhance polymer deposition. Anodized aluminium electrodes, however, are not sacrificial and the surfaces are insulators which will change the ionisation conditions, Ref. (12) did not find polymer formation on anodized electrodes, also ref. 12 reported more C_2Cl_4 production (using GLC analysis) for stainless steel electrodes than for anodized aluminium.

14. J.L. Vossen, J. Electrochem. Soc., SSST, Vol. 126, No.2, Feb. 1979, p.322.

15. R.A. Morgan, Vacuum (May) Vol. 32, No.5, p.297, 1982.

16. R.L. Bersin, Solid State Tech., April 1978.

17. R.L. Maddox and H.L. Parker, Solid State Tech., April 1978, p.107.

18. R.C. Booth and C.J. Heslop, Thin Solid Films, Vol. 65, p.111, 1979.

19. S.C. Brown, Basic Data of Plasma Physics, MIT Press, Cambridge, Mass., 1966.

20. G.C. Schwartz and P.M. Schaible, Solid State Tech., Nov. 1980, p.85.

21. Kalter and Van DeVen, J. Electrochem. Extended Abstr. No.76, p.335, 1976.

22. Jinno and Kinoshita and Matsumato, J. Electrochem. Soc., No.124, p.1528, 1977.

23. A.R. Reinberg, "Etching for pattern definition" (Hughes and Rand, eds.) p.91, Electrochem. Soc., Princeton, New Jersey, 1976.

24. C.J. Mogab and T.A. Shankoff, J. Electrochem. Soc., Vol.124, p.1766, 1977.

25. Flamm et al., J. Appl. Phys., Vol. 52, p.3633, 1981.

26. A. Jacob, Solid State Tech., June 1977, p.31.

27. B.N. Chapman and V.J. Minkiewicz, J. Vac. Sci. and Tech., Vol. 15, No.2, March/April 1978.

28. B.N. Chapman, "Glow Discharge Process", 1979, Wiley, New York.

29. J.L. Mauer and Joseph S. Logen, J. Vac. Sci. and Tech., Vol. 16, No.2, 1979, p.404.

30. J.L. Mauer, J.S. Logan, L.B. Zielinst and G.C. Schwartz, J. Vac. Sci. and Tech., Vol. 15, No.5, Sept./Oct. 1978, p.1734.

31. J.D. Chinn, I. Adeside, E.D. Wolf and R.C. Tiberio, J. Vac. Sci. and Tech., Vol. 19, No.4, Nov./Dec. 81, p.1418.

32. J.W. Coburn and E. Kay, J. Appl. Phys. Vol. 43, No.12, Dec. 1972, p.4965.

33. L.M. Ephrath and D.J. Dimaria, Solid State Tech., April 1981,p.182.

34. P.M. Schaible, N.C. Metzger and J.P. Anderson, J. Vac. Sci. and Tech., Vol. 15, No.2, March/April 1978.

35. N.K. Wang, D. Mayden, H.J. Levinstein, Solid State Tech., Aug. 1980, p.122.

36. L.M. Ephrath, IEEE 1980, p.402, CH1616-2/80/0000-0402.

37. Ref.(1) cited in D. Bollinger and R. Fink, Solid State Tech., Nov. 1980, p.79.

38. H. Boyd and M.S. Tank, Solid State Tech., April 1979, p.133.

39. D.F. Downey, W.R. Bottoms and P.R. Hanley, Solid State Tech., Feb. 1981, p.121.

40. B.A. Heath, Solid State Tech., Oct. 1981, p.75.

41. N. Hoskawa, R. Matsuzaki and T. Asamaki, Jap. J. Appl. Phys. Supp. 2, Part 1, 1974, p.435.

42. W.R. Harshbarger and R.A. Porter, Solid State Tech., April 1978, p.99.

43. C.J. Mogab, A.C. Adams and D.F. Flamm, J. Appl. Phys., Vol. 49, (7) p.3796.

44. F. Kaufman and J.R. Kelso, J. Chem. Phys. 32, p.301, 1960.

45. M.J. Vasile, J. Appl. Phys. Vol. 51 (5), May 1980, p.2503.

46. Y. Horiike and M. Shibagki, Jap. J. Appl. Phys., Vol. 15, 13, 1976 Supp.

47. D.L. Flamm, Solid State Tech., April 1979, p.109.

48. Ref. 17 sited in ref. 47.

49. D.L. Flamm, to be published.

50. R.A. Heineke, Solid State Elec., Vol. 1, p.1039-1040, 1976.

51. C.I.M. Beenakker, J.H.S. Van Dommelen and R.P.J. Van de Call, J. Appl. Phys. Vol. 52, No.1, 1981.

52. G. Somolinsky and D.L. Flamm, J. Appl. Phys. Vol. 50, No.7, July 1979, p.4982.

53. R. Agostino, F. Cramarossa, S. De Bendocts and G. Ferraro, J. Appl. Phys. Vol. 52, No.3, March 1982, p.1259.

54. R. Heineke, internal report.

55. J.W. Coburn and E. Kay, IBM J. Res. Dev., Vol. 23, No.1, Jan. 1979, p.33.

56. H.W. Lehmann and R. Widmer, Appl. Phys. Lett. 32 (3), Feb. 1978.

57. Curine and Lang, Surface Science 64, 209, 1977.

58. S. Matsuo, J. Vac. Sci. and Tech., Vol. 17, No.2, Mar./April 1980, p.587.

59. J.W. Coburn, H.F. Winters and T.J. Chang, J. Appl. Phys. Vol. 48, No.8, 1977.

60. R. d'Agostino and D.L. Flamm, J. Appl. Phys. Vol. 52, No.1, Jan. 1981, p.162.

61. K.M. Elisele, J. Electrochem. Soc. (SSS&T) Vol. 128, No.7, Jan. 1981, p.123.

62. C.J. Mogab and D.L. Flamm, to be published.

63. J.W. Coburn and H.F. Winters, J. Vac. Sci. and Tech., Vol.16, No.2, 1979.

64. L. Holland, L. Laurenson, R.E. Hurley and K. Williams, Nuclear Instruments and Methods 111, 1973, p.555.

65. H.F. Winters, J.W. Coburn and E. Kay, J. Appl. Phys. Vol. 48, No.12, 1977, p.4973.

66. Private communication, M. Gill.

67. R.A. Morgan, Work in BT to be reported.

68. H. Toyoda, H. Itakura and H. Komiya, Japn. J. Appl. Phys. Vol. 51, No.5, May 1980, p.2909.

69. T.M. Mayer, R.A. Baker and L.T. Whitman, J. Vac. Sci. and Tech., Vol. 18, No.2, March 1981.

70. H.F. Winters, J. Appl. Phys. Vol. 49, No.10, Oct. 1978, p.5165.

71. E.A. Truesdale, G. Smohinsky and T.M. Mayer, J. Appl. Phys., Vol. 51, No.5, May 1980, p.2909.

72. D.L. Flamm, V.M. Donnelly and J.A. Mayer, J. Appl. Phys., Vol. 5, No.2, May 1981.

73. M. Sato and H. Nakamura, J. Vac. Sci. and Tech., Vol. 20, No.2, Feb. 1982, p.163.

74. H.W. Lehman and R. Widmer, J. Vac. Sci. and Tech., Vol.15, No.2, 1978, p.319.

75. S. Thomas, J. Appl. Phys., Vol. 45, p.161, 1974.

76. M.J. Vasile, J. Appl. Phys. Vol.51, No.5, May 1980, p.2510.

77. M. Gill, Solid State Elec., Vol. 23, 1980, p.995.

78. G.C. Schwartz, private communication.

79. P.D. Parry and A.F. Roddle, Solid State Elec., April 1979, p.125.

80. S. Matsuo, Appl. Phys. Lett., p.369, No.1, May 1980.

81. D.L. Flamm, P.L. Cowan and J.A. Golovchenko, J. Vac. Sci. and Tech., Vol.17, No.6, 1980, p.1341.

82. C.I.M. Beenaker, R.P.J. Van de Poll and J. Dielman, Abstr. ECS, 256 Fall Meeting, Denver, Colorado, U.S.A. Oct. 11-16, 1981.

83. D.L. Flamm, J. Appl. Phys., Vol. 51, No.11, 1980, p.5688.

84. J. Paraszczak and M. Hatzakis, J. Vac. Sci. and Tech., Vol.19, No.4, p.886-91, 1981.

85. V.M. Donnelly and D.L. Flamm, Solid State Tech., April 1981, p.161.

86. G.C. Schwartz and P.M. Schaible, J. Vac. Science and Tech., Vol.16, No.2, 1979, p.410.

87. W.W. Yao and R.H. Bruce, Abstr. 268, Denver, Colorado, Oct. 1981.

88. K. Zhang and L. Yeh, J. Vac. Sci. and Tech., Vol.19, No.3, Sept./Oct. 1981, p.628.

89. Cotton and Wilkinson, "Advanced Inorganic Chemistry", Second Edition, Wiley Interscience, New York, 1966.

90. Berezhnoi, "Silicon and its Binary Systems", Consultants Bureau, New York, 1960, p.39.

91. S.H. Kravitz, B.R. Soller and D.J. Rieger, ECS Meeting, Denver, Colorado, Oct. 1981.

92. G.C. Schwartz and P.M. Scharible, Plasma USA 80.

93. R.G. Poulsen, H. Nentwich and S. Ingrey, International Electron Devices Meeting, Technical Digest, Washington DC, USA, Dec. 76, p.205-206.

94. P.M. Schaible and G.C. Schwartz, J. Vac. Sci. and Tech., Vol.16, No.2, Mar./April, 1979, p.377.

95. B.J. Curtis and H.R. Brunner, 5th International Symposium on Plasma Chemistry, Heriot-Watt University, Edinburgh, August 1981, p.318, edited by B. Waldie and G.A. Farnell.

96. K. Tokunaga, F.C. Redeker, D.A. Danner and D.W. Hess, J. Electrochem. Soc., April 1981, p.851.

97. M. Sato and H. Nakamura, J. Vac. Sci. and Tech., Vol.20, No.2, Feb. 1982, p.186.

98. R. Walter, one day conference, Feb. 1982 at the IOP.

99. Physics of Semiconductor Devices, SM SZE.

100. Casey and Panisch, Hetrojunction Lasers.

101. G. Smolinsky, R.P. Chang and T.M. Mayer, J. Vac. Sci. and Tech., Vol. 18, No. 1, Jan./Feb. 1981, p.12.

102. D. Asaro et al., International Symposium, Vienna, Austria, Sept. 22, 1980.

103. Briton et al., Applied Physics Lett. 37, 1980, p.411.

104. L.A. Caldren and J.A. Rentschler, J. Vac. Sci. and Tech., Vol.19, No. 2, 1981, p.225.

105. R.P.H. Chang and S. Darack, Appl. Phys. Lett. 38(11), 1 June, 1981, p.898.

106. D.E. Iobboston, V.H. Donnelly, D.L. Flamm and B.S. Duncan, The Electrochem. Soc., 160th Fall Meeting, Denver, Colorado, U.S.A., October 11-16, 1981, Abstr. No.276.

107. V.M. Donnelly, D.L. Flamm and G.J. Collins, The Electrochem. Soc. 160th Fall Meeting, Denver, Colorado, USA, Oct. 11-16, 1981, Abstr. No.257.

108. H.R. Koeing and L.I. Maissel, IBM J. Res. Dev. No.14, March 1970, p.168.

109. J.S. Logan, IBM J. Res. Dev. Vol.14, p.172, 1970.

110. J.S. Logan, J.H. Keller and R.G. Simmons, J. Vac. Sci. and Tech., Vol.6, p.120, 1969.

111. H. Norstrom, Vacuum, Vol.29, No.10, p.341, 1979.

112. H. Norstrom, Vacuum, Vol.29, No.11, p.443, 1979.

113. R.H. Bruce and A.R. Reinberg, Internat. Conf. on Microlithography, Sept. 30-Oct. 2, 180, Amsterdam, Proc. Microcircuit Engineering 1980, Delft University Press, p.533.

114. G. Francis, Ionization phenomena in Gases, Butterworth, London, p.148 (1st edition) (1960).

115. N.A. Popor and N.A. Kapstor, J. Exp. Theor. Phys. 30, 1956, p.68.

116. R.H. Bruce, Proc. of Symp. on Plasma Etching and Deposition, Vol. 81-1, p.243, Electrochem. Soc. 1981.

117. D.L. Flamm, V.M. Donnolly, R.H. Bruce and R.G. Collins, 5th International Symposium on Plasma Chemistry, Heriot-Watt University, Edinburgh, August 1981, p.307, edited by B. Waldie and G.A. Farnell.

118. M. St. J. Burden and K.B. Cross, Vacuum, Vol.29, No.1, 1979.

119. J. Maleham and N.P. Armstrong, Colloquium on Dry Etching at the IEE, Savoy Place, London, March 2nd, 1982.

120. A.D. MacDonald and S.J. Tetenbaum, Gaseous Electronics, Vol.1, ed. Merle N. Hirsh and H.J. Oskam, Academic Press, New York and London, 1978.

121. L. Holland, W. Steckelmacher and J. Yarwood, eds. Vacuum Manual, Spon, London 1974.

122. E.W.B. Gill and A. von Engel, Proc. Roy. Soc. A192 (1948), 446-63.

123. J.M. Keller and W.B. Pennebaker, IBM J. Res. Dev. 23 (1), 1979, p.3.

124. "Techniques and Applications of Plasma Chemistry", edited by J.R. Hollahan and A.T. Bell, Wiley Interscience, 1974, New York.

125. Molecular Spectroscopy, J. Spice, 2nd Edition, McGraw, 1964.

126. J. Polman, Physica, Neth., 34, 1967, p.310-16.

127. S. Vacquie, J. Bacri and G. Serrot, Comp. Rend. 266 (1968), p.387-390.

128. As the operating voltage increases the dark space width will be larger (Childs law) but as the pressure increases voltage will reduce and λ_e, λ_D.

129. M. Gill, Plasma Deposition Meeting at the IEE Savoy Place, London, 3rd March 1982.

130. R.A. Morgan, to be published.

131. R.H. Bruce, Solid State Tech., Oct. 1981, p.64.

132. M. Gill, Research Memoranda No.R2/020/82.

133. C. Dobson and P. Granger, Electrotech. Equipments, Bristol, England.

134. H. Mott-Smith and I. Langmuir, Physical Rev. 28 (1926), 727.

135. R.M. Clements, J. Vac. Sci. Tech. 15 (1978), 193.

136. J.A. Thornton, J. Vac. Sci. Tech. 15 (1978), 188.

137. G. Laframboise, University of Toronto, Institute for Avo Space Studies, Report No.100 (1966).

138. E.O. Johnson and L. Malter, Phys. Rev. 80 (1950), 58.

139. H.W. Rundle, D.R. Clark and J.M. Deckers, Can. J. Phys. 51 (1973), 144.

140. G.A. Woolsey, I.C. Plumb and D.B. Lewis, J. Phys. D: Appl. Phys. 6 (1973), 1883.

141. C.V. Goodall and D. Smith, Plasma Phys. 10, 1968, 249.

142. D. Smith and I.C. Plumb, J. Phys. D: Appl. Phys. 6 (1973), 196.

143. E. Eser and R.E. Ogilvie, J. Vac. Sci. Tech. 15, (1978), 199.

144. A. Cantin and R.R.J. Gagne, Final Report, Can. Def. Res. Board Grant No.9510-63.

145. D. Bohm,"The Characteristics of Electrical Discharges in Magnetic Fields", ed. A. Guthrie and R.K. Wakerling, McGraw Hill, New York and London, 1949.

146. E.H. Hirsch, Int. J. Elec. 19 (1965), 537.

147. J.F. Waymouth, J. Appl. Phys. 30 (1959), 1404.

148. R.L.F. Boyd and N.D. Twiddy, Proc. Roy. Soc. (London) A 250 (1959), 23.

149. E.E. Mainwaring and J.D. Swift, J. Phys. D: Appl. Phys. 5 (1972), 1433.

150. A. Cantin and R.R.J. Cagne, Appl. Phys. Letts. 30 (1977), 316.

151. M.J. Kushner, J. Appl. Phys. 53 (1982), 2939.

152. L. Holland, W. Steckelmacher and J. Yarwood, Vacuum Manual E&FN spon, London, p.368 (1974).

153. W.R. Larsen, C.W. Wilmsen and P.W. Chan, J. Appl. Phys. 44 (1973), 2153.

154. J. Taillet, Journal de Physique, Letters (Paris) 40, L223, 1979.

155. L. Holland, Vacuum 28 (1978), 434.

156. G. Francis, Ionization Phenomena in Gases, Butterworths, London,p.148 (1st edition) 1960.

157. M.J. Bowden, Solid State Tech., June 1981.

158. E.A. Chandross, E. Reichmanis, C.W. Wilkins and R.L. Hartless, Solid State Tech., August 1981.

159. Hunt Technical Bulletin, No.41 and 42.

160. H. Namatsu, Y. Ozaki and K. Hirata, J. Vac. Sci. Tech. 21 (1982), 672.

161. J. Pacansky and J.R. Lyeria, IBM J. Res. Develop. 23, No.1, Jan. 1979.

162. R.A. Morgan, Laboratory Report R2.4.3/02.

163. H. Hiraoka and J. Pacansky, J. Vac. Sci. Technol. 19 (1981), 1132.

164. R. McMahon, H. Ahmed, R. Dobson and J.D. Speight, Electronic Letts. 10 (1980), 295.

165. J. Brandrup and E. Immergut, Polymer Handbook (Wiley Interscience, New York).

166. K. Harada, J. Appl. Polymer Sci. 26 (1981), 3395.

167. Y. Yamashita, K. Ogura, M. Kunishi, R. Kawazu, S. Ohno and Y. Mizokami, J. Vac. Sci. Tech. 16 (1979), 2026.

168. E.A. Chardross et al., Solid State Tech., August 1981.

169. S.M. Ojha, Vacuum 27 (1977), 65.

170. J.L. Vossen, J.J. O'Neill, Jr., G.W. Hughes, F.A. Taft and R. Snedter, J. Vac. Sci. Tech. 17 (1980), 400.

171. L. Holland and C.R.D. Prestland, Vacuum 22 (1972), 133-149.

172. L. Holland, Thin Solid Films 86 (1981), 227-239.

173. Photon Acoustic Spectroscopy developed at Salford University by M. Slifkin.

174. L.F. Thompson, C.G. Wilson and M.J. Bowden, "Introduction to Microlithography", ACS Symposium Series 219, p.93, 1983.

SUBJECT INDEX